Sequences

A Self-Study Guide to Mathematics

Volume 2

First Edition

Jianlun Xu

Printed by CreateSpace, An Amazon.com Company

ISBN-13:978-1548279912
ISBN-10:1548279919

Preface

This book is written for high school students, college students, and anyone interested in teaching themselves math. The goal of the book is to help you establish a solid foundation of mathematics for your advanced studies and the preparation of SAT and ACT.

The book is one of a set of books, *"A Self-Study Guide to Mathematics"*. Each book covers one particular subject of high school or college mathematics. The characteristics of this book include

- Extensive coverage of a particular math subject
- Plain language to explain math concept
- Plenty of math proofs and explanations to tell you why
- Abundant detailed examples to tell you how-to step by step

Someone may say "I know what I learned from my math class but I don't know why in detail and how to solve new math problems step by step." If this sounds like you this book is for you.

Some students feel mathematics boring because they may not be trained to think in a mathematical way during their studies. When you complete this book, you will be gradually trained to think logically through plenty of proofs and examples in detail. You will find that math likes a fun game. That reminds me of a dialogue between my daughter and I. When she was in middle school she played piano and also took part in math contests. One day she asked me "Do you think that someone good at piano will also be good at math?" "Why do you ask such question?" I wondered. "In my class a boy is good at math and competes against me in a math contest. He plays piano very well too". "That is quite possible, math is harmonious like music". I answered. Yes, math is beautiful.

The set of books, *"A Self-Study Guide to Mathematics"*, is your math tutor at home. Good luck in your math study and the preparation of SAT and ACT.

I appreciate the support from my wife and daughter who make this book possible.

Jianlun Xu

Contents

4   Recursive Sequences

Type 1: $a_{n+1} = a_n + f(n)$

Type 2: $a_{n+1} = f(n)a_n$

Type 3: $a_{n+1} = pa_n + q$

Type 4: $a_{n+1} = pa_n + q(n)$

Type 5: $a_{n+1} = pa_n + qa_{n-1}$

Type 6: $a_{n+1} = pa_n + qa_{n-1} + c$

5   Arithmetic Sequences of Higher Order

 # Essentials of Sequences

▌ 1.1 Introduction

Sequences is a very important aspect of high school mathematics. It is the foundation of the limits in calculus. A **sequence** is a collection of infinite number of numbers listed in a certain order. For each positive integer n ($n \in N^+$) there is a corresponding term a_n in the sequence. For instance,

1) All even numbers $\qquad\qquad\qquad\qquad$ *2, 4, 6, 8, 10, ⋯, 2n, ⋯*

2) Infinite number of the constants $\qquad$ *8, 8, 8, 8, 8, ⋯*

3) ⬤　⬤⬤　⬤⬤⬤　⬤⬤⬤⬤ ⋯ $\qquad$ *1, 3, 6, 10, ⋯, (1 + 2 + ... + n),*

Integer n	1	2	3	4	...	n	...
Term a_n	1	1+2=3	1+2+3=6	1+2+3+4=10	...	1+2+ ... +(n–1) + n	...

4) All approximate values of $\sqrt{2}$ $\qquad$ *1, 1.4, 1.41, 1.414, ⋯*

5) $f(n) = (-1)^{n+1} 4n$, (n = 1, 2, 3, ⋯) $\qquad$ $4, -8, 12, -16 ⋯, (-1)^{n+1} 4n, ⋯$

DEFINITION   Sequence

A sequence of numbers is a discrete function, whose domain is a set of natural numbers or a subset of natural numbers. The values

$$f(1), f(2), f(3), \cdots, f(n), \cdots \qquad\qquad (n \in N^+)$$

are called the terms of the sequence and are listed in certain order.

Sequences fall into two categories
- **Infinite sequence**　A sequence which has infinite number of terms.
- **Finite sequence**　A sequence which has finite number of terms.

Generally a sequence is refereed to as an infinite sequence unless it is finite in nature or specified as a finite sequence. A sequence can be expressed as

$$\{a_1, a_2, \cdots, a_n, \cdots \}.$$

There is a corresponding term for each positive integer n such as the first term a_1, the second term a_2, and so on. The nth **term**, denoted a_n, is called the **general term** of a sequence. The subscript n is a set of consecutive positive integers, which acts as the index of the terms of a sequence.

■ 1.2 Notations of Sequences

Because a sequence can be determined by its general term, $a_n = f(n)$, the symbol $\{a_n\}$ is used to denote a sequence $\{a_1, a_2, \cdots a_n, \cdots\}$.

The expression $\{a_n\} : \{x_1, x_2, \cdots x_n, \cdots\}$ means "Let $\{a_n\}$ be $\{x_1, x_2, \cdots x_n, \cdots\}$."

The following points will help you to understand notation of sequences better.

- The symbol a_n denotes the nth term of a sequence $\{a_n\}$ therefore the symbols $\{a_n\}$ and a_n have different meaning.
- Simply listing the first few terms without the formula of general term, $a_n = f(n)$, can **not** guarantee the uniqueness of a sequence. The formula of the general term, $a_n = f(n)$, inducted from the first few terms should be suitable for all rest terms.
- The formula of the general term of a sequence may not be unique.
- A sequence may not have any expressible formula of its general term a_n.
- The index n is a natural number ($n \in N^+$) to indicate the position of a term in a sequence. It does not have to begin at 1.

We have the following notations to represent a sequence such as

- Function notation,
- Subscript notation,
- Recursive notation, and
- Summation notation.

▶ Function Notation

With function notation, the general term a_n of a sequence $\{a_n\}$ is denoted by function

$$a_n = f(n) \quad (n \in N^+)$$

Then the sequence can be denoted by $\{f(n)\}$. The index n is an independent variable and each term of the sequence is a solution of the function $f(n)$.

▶ Subscript Notation

Simply list the first few terms a_1, a_2, $\cdots$, and the formula of the general term a_n like

$$\{a_1, a_2, \cdots f(n), \cdots\}$$

Example 1.2.1   *Function Notation*

Write out the notation of the following sequences by function notation and subscription notation.

1) $a_n = \sqrt{2n+1}$ 2) $a_n = (-1)^n + 3n$ 3) $a_n = \dfrac{3n^2 - 1}{n+3}$

Solution:

1) $\{\sqrt{2n+1}\}$ and $\{\sqrt{3},\ \sqrt{5},\ \sqrt{7},\ ...,\ \sqrt{2n+1},\ ...\}$

2) $\{(-1)^n + 3n\}$ and $\{2,\ 7,\ 8,\ ...,\ (-1)^n + 3n,\ ...\}$

3) $\{\dfrac{3n^2 - 1}{n+3}\}$ and $\{\dfrac{1}{2},\ \dfrac{11}{5},\ \dfrac{26}{6},\ ...,\ \dfrac{3n^2 - 1}{n+3},\ ...\}$

Example 1.2.2   *Find the General Term of a Sequence*

Count the smallest squares in each graph to construct the sequence $\{a_n\}$. Write out the first four terms of the sequence and its general term a_n.

Solution:

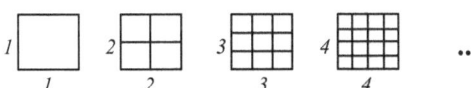

When $i = 1, 2, 3, 4$, the ith graph contains $1, 4, 9$, and 16 squares respectively. We can write the first four terms like $\{1, 4, 9, 16\}$.

The nth graph contains $n \cdot n$ squares thus

$$a_n = n^2.$$

Therefore the complete sequence becomes

$$\{a_n\} : \{1, 4, 9, 16, \cdots, n^2, \cdots\}.$$

Example 1.2.3   *Find the General Term of a Sequence*

Count the smallest triangles in each graph to construct the sequence $\{a_n\}$. Write out the first four terms of the sequence and its general term a_n.

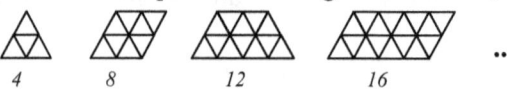

4   8   12   16   ···

Solution:

We will figure out the pattern forming the graph.

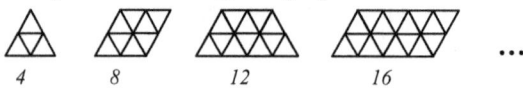

4   8   12   16   ···

When $i = 1, 2, 3, 4$, the ith graph contains *4, 8, 12, 16* triangles respectively. We can write the first four terms like $\{4, 8, 12, 16\}$. The nth graph contains $4n$ triangles.

Let $a_n = 4n$.

Then we can write the complete sequence

$$\{a_n\}:\{4, 8, 12, 16, \cdots, 4n, \cdots\}.$$

Example 1.2.4   *Find the General Term of a Sequence*

If the first few terms of the following sequences are given, find the general term a_n.

1) $\{2, 4, 8, 16, \cdots\}$
2) $\{-3, 6, -9, 12, -15, \cdots\}$.
3) $\{3, -6, 9, -12, 15, \cdots\}$.
4) $\{\dfrac{1}{2}, \dfrac{2}{3}, \dfrac{3}{4}, \dfrac{4}{5}, \dfrac{5}{6}, \cdots\}$.
5) $\{1, 1, 2, 2, 3, 3, 4, 4, \cdots\}$
6) $\{3, 7, 15, 31, 63, \cdots\}$.
7) $\{1, 2, 3, 4, 5, 6, 7, \cdots\}$.

Solution:

1) $\{a_n\}:\{2, 4, 8, 16, \cdots\}$.   : $a_n = 2^n$

2) $\{a_n\}:\{-3, 6, -9, 12, -15, \cdots\}$.   : $a_n = (-1)^n\, 3n$

3) $\{a_n\}:\{3, -6, 9, -12, 15, \cdots\}$.   : $a_n = (-1)^{n+1}\, 3n$

4) $\{a_n\}:\{\dfrac{1}{2}, \dfrac{2}{3}, \dfrac{3}{4}, \dfrac{4}{5}, \dfrac{5}{6}, \cdots\}$.   : $a_n = \dfrac{n}{n+1}$

5) $\{a_n\}:\{1, 1, 2, 2, 3, 3, 4, 4, \cdots\}$: $a_n = 2^n$

6) $\{a_n\}:\{3, 7, 15, 31, 63, \cdots\}$.   : $a_1 = 3, a_n = 2a_{n-1} + 1 \ (n \geqslant 2)$

7) $\{a_n\}:\{1, 2, 3, 4, 5, 6, 7, \cdots\}$.   : $a_1 = 1, a_2 = 2, a_n = 2a_{n-1} - a_{n-2} \ (n \geqslant 3)$

Example 1.2.5   *Find the General Term of a Sequence*

If the first few terms of the following sequences are given, find the general term.

1) $\{\frac{1}{1\cdot 2}, \frac{1}{2\cdot 3}, \frac{1}{3\cdot 4}, \frac{1}{4\cdot 5}, \cdots \}$

2) $\{1, 2, -7, 14, -23, 34, \cdots \}$

3) $\{9, 99, 999, 9999, \cdots \}$

4) $\{5, 8, 11, 14, 17, \cdots \}$

5) $\{1, -1, 1, -1, 1, \cdots \}$

6) $\{-3, 9, -27, 81, -243, \cdots \}$

Solution:

We list the first few terms with their index n, then find out the relationship between index n and general term a_n and fill it in the column of index n.

1)

Index n	1	2	3	4	...	n	...
a_n	$\frac{1}{1\cdot 2}$	$\frac{1}{2\cdot 3}$	$\frac{1}{3\cdot 4}$	$\frac{1}{4\cdot 5}$	...	$a_n = \frac{1}{n(n+1)}$	..

For each term, the denominator is the multiply of its index n and next index $n+1$ then the general term is $a_n = \frac{1}{n\cdot(n+1)}$.

2)

Index n	1	2	3	4	5	6	...	n	...
a_n	n/d	2	-7	14	-23	34	...	$a_n=(-1)^n(n^2-2)$ $(n\geqslant 2)$	...

Starting from $n = 2$, the absolute value of the general term is $|n^2-2|$ and changes its sign alternatively. The general term is $a_n=(-1)^n(n^2-2)$ $(n\geqslant 2)$. (n/d = "not defined")

3)

Index n	1	2	3	4	...	n	...
a_n	9	99	999	9999	...	$a_n=10^n-1$	...

Since $9 = 10^1-1$, $99 = 10^2 -1$, $999 =10^3 - $, $\cdots$ we have the general term $a_n=10^n-1$.

4)

Index n	1	2	3	4	5	...	n	...
a_n	5	8	11	14	17	...	$a_n=3n+2$	...

Since $5=3\cdot 1+2$, $8=3\cdot 2+2$, $11=3\cdot 3+2$, $\cdots$, we have the general term $a_n=3n+2$.

5)

Index n	1	2	3	4	...	n	...
a_n	1	-1	1	-1	...	$a_n=(-1)^{n+1}$	...
a_n	1	-1	1	-1	...	$a_n=\cos n\pi$	...

Since the sign of the general term a_n changes alternativelyand its absolute value is 1, we have $a_n=(-1)^{n+1}$ or $a_n=\cos n\pi$. In this case the general term is not unique.

6)

Index n	1	2	3	4	5	...	n	...
a_n	-3	9	-27	81	-243	...	$(-1)^n 3^n$	...

$(-1)^n$ is used to express the change of the sign of term a_n.

▶ Recursive Notation

If any term of a sequence can be expressed by its one or more previous terms it is called **recursive sequence**. Thus a recursive sequences can be defined by
*   The first k terms given $(a_1, a_2, \cdots, a_k)$, which are called the initials, and
*   The formula defining the recursive relationship between the general term a_n and its previous k terms $a_{n-1}, a_{n-2}, \cdots, a_{n-k}$.

Then $$a_n = f(a_{n-1}, a_{n-2}, \cdots, a_{n-k}) \qquad (1.2.1)$$
A recursive sequence can be defined by the notion above.

Example 1.2.6   *The First Five Terms of a Recursive Sequence*
Find the first five terms of the following recursive sequences.

1) $a_1=1, a_n=a_{n-1}+n \ (n \geqslant 2)$,   2) $a_3=4, a_4=2, a_n=\dfrac{2a_{n-1}}{a_{n-2}} \ (n \geqslant 5)$

3) The **Fibonacci** sequence $a_1=1, a_2=1,$ and $a_n=a_{n-1}+a_{n-2}. \ (n \geqslant 3)$.

Solution:

$$
1) \begin{cases} a_1=1 \\ a_2=a_1+2=3 \\ a_3=a_2+3=6 \\ a_4=a_3+4=10 \\ \cdots \end{cases}
\quad
2) \begin{cases} a_3=4 \\ a_4=2 \\ a_5=1 \\ a_6=2a_4+a_5=14 \\ \cdots \end{cases}
\quad
3) \begin{cases} a_1=1 \\ a_2=1 \\ a_3=a_2+a_1=2 \\ a_4=a_3+a_2=3 \\ \cdots \end{cases}
$$

$\{a_n\}:\{1, 3, 6, 10, \cdots\}$ $\{a_n\}:\{4, 2, 10, 14, \cdots\}$ $\{a_n\}:\{1, 1, 2, 3, 5, 8, \cdots\}$

Example 1.2.7   *The General Term of a Recursive Sequence*

If $\{a_n\}$ is a recursive sequence, $a_n = 3 a_{n-1} + 4$ and $a_1 = 1$ ($n \geqslant 1$), find a_n.

Solution:

Arrange $a_n = 3 a_{n-1} + 4$ to $a_n + 2 = 3(a_{n-1} + 2)$.

We have $n-1$ equations

$$\begin{cases} a_n + 2 = 3(a_{n-1} + 2) \\ a_{n-1} + 2 = 3(a_{n-2} + 2) \\ \cdots \\ a_2 + 2 = 3(a_1 + 2) \end{cases}$$

Multiply these $n-1$ formulas we obtain

$$a_n + 2 = 3^{n-1}(a_1 + 2)$$

Since $a_1 = 1$ $a_n = 3^n - 2.$

▶ Summation Notation

The sum of the first n terms of an infinite sequence $\{a_n\}$ is defined as

$$S_n = a_1 + a_2 + a_3 + \cdots + a_n = \sum_{i=1}^{n} a_i \tag{1.2.2}$$

Where n is a finite natural number.

Since $a_1 = S_1, a_2 = S_2 - S_1, \cdots a_n = S_n - S_{n-1}, \cdots,$

$$\{a_n\} = \{S_1, (S_2 - S_1), \cdots (S_n - S_{n-1}), \cdots\}.$$

The general term of a sequence $\{a_n\}$ can be determined by the following formula.

$$a_n = S_n - S_{n-1} \qquad (n \geqslant 2) \tag{1.2.3}$$

In this way a unique sequence $\{a_n\}$ is defined. It is called **summation notation**.

Because the formula (1.2.3) is valid when $n \geqslant 2$ it is necessary to check the first term a_1. If a_1 is suitable for the formula the general term a_n can be expressed as a single formula

$$a_n = S_n - S_{n-1} \qquad (\forall n) \tag{1.2.4}$$

else a_n can be expressed by a piecewise formula.

$$a_n = \begin{cases} S_1 & (n=1) \\ S_n - S_{n-1} & (n \geqslant 2) \end{cases} \tag{1.2.5}$$

Example 1.2.8   *Summation of Sequences*

Find the following sums.

1) $\sum\limits_{i=2}^{5}(2i+1)$ 2) $\sum\limits_{i=1}^{5}\dfrac{i-1}{i+1}$ 3) $\sum\limits_{i=1}^{6}(-1)^{i}(2i+1)$

Solution:

1) $\sum\limits_{i=1}^{5}(2i+1)=(2\cdot1+1)+(2\cdot2+1)+(2\cdot3+1)+(2\cdot4+1)+(2\cdot5+1)=35$

2) $\sum\limits_{i=2}^{5}\dfrac{i-1}{i+1}=\dfrac{2-1}{2+1}+\dfrac{3-1}{3+1}+\dfrac{4-1}{4+1}+\dfrac{5-1}{5+1}=\dfrac{1}{3}+\dfrac{2}{4}+\dfrac{3}{5}+\dfrac{4}{6}=\dfrac{21}{10}$

3) $\sum\limits_{i=1}^{6}(-1)^{i}(2i+1)=(-1)^{1}(2\cdot1+1)+(-1)^{2}(2\cdot2+1)+(-1)^{3}(2\cdot3+1)=6$

Example 1.2.9   *Summation Notation*

If $S_{n}=3n^{2}+2n+1$, find general term a_{n} of the sequence $\{a_{n}\}$.

Solution:

The first term a_{1}

$$a_{1}=S_{1}=3\cdot1^{2}+2\cdot1+1=6. \qquad (n=1)$$

The general term a_{n}

$$S_{n-1}=3\cdot(n-1)^{2}+2\cdot(n-1)+1 \qquad (n\geqslant2)$$
$$=3n^{2}-4n+2$$

Then

$$a_{n}=S_{n}-S_{n-1}=6n-1.$$

As a_{1} is not suitable for the formula above. The general term a_{n} is defined by the piecewise formula below.

$$a_{n}=\begin{cases}6 & (n=1)\\6n-1 & (n\geqslant2)\end{cases}$$

▌1.3 Types of Sequences

Infinite Sequences
An infinite sequence is the sequence whose number of terms is infinite.
$$1, 2, 3, 4, \cdots, n, \cdots$$

Finite Sequences
A finite sequence consists of the first n terms of an infinite sequence, n is a finite number.
$$1, 2, 3, 4, \cdots, n$$

Constant Sequences
A sequence is a constant sequence if $a_{n+1} = a_n$ ($\forall n$).
$$8, 8, 8, \cdots, 8, \cdots$$

Increasing Sequences
A sequence is an increasing sequence if $a_n < a_{n+1}$ ($\forall n$).
$$2, 4, 6, 8, \cdots$$

Decreasing Sequences
A sequence is a decreasing sequence if $a_n > a_{n+1}$ ($\forall n$).
$$9, 7, 5, 3, \cdots$$

Monotone Sequences
A sequence is a monotone sequence if it is either increasing sequence or decreasing sequence.

Non-decreasing Sequences
A sequence $\{a_n\}$ is a non-decreasing sequence if
$$a_n \leqslant a_{n+1} \quad (\forall n)$$

Non-increasing Sequences
A sequence $\{a_n\}$ is a non-increasing sequence if
$$a_n \geqslant a_{n+1} \quad (\forall n)$$

Oscillating Sequences
A sequence $\{a_n\}$ is an oscillating sequence if $a_n < a_{n+1}$ and $a_n > a_{n+1}$ alternatively.
$$1, -1, 1, -1, 1, \cdots$$

Bounded Sequences
A sequence $\{a_n\}$ is a bounded sequence if there exists a positive real number M to let the absolute value of any term of the sequence less than or equal to M:
$$|a_n| \leqslant M \quad \forall n \geqslant 1$$
All finite sequences are bounded sequences but an infinite sequence may or may not be a bounded sequence.

Arithmetic Sequences
A sequence is a arithmetic sequence if the differences between any two consecutive terms are the same.
$$3, 6, 9, 12, 15, \cdots, 3n,$$

Geometric Sequences
A sequence is a geometric sequence if the ratios between any two consecutive terms are the same.
$$3, 9, 27, 81, \cdots, 3^n$$

Example 1.3.1   *Types of Sequences*

Identify the type of the following sequences.

1) $a_n = \dfrac{n}{n+1}$ 2) $a_n = 10 - 2n$ 3) $a_n = \cos n\pi$

4) $a_n = 3^n$ 5) $a_n = \sin n\pi$ $(1 \leqslant n \leqslant 10000)$ 6) $1, 1^2, 1^3, \cdots, 1^{100}$

Solution:

1) $a_n = \dfrac{n}{n+1}$

- $\{a_n\}$ is not specified as a finite sequence then it is consider as an infinite sequence.

- Sine $a_{n+1} - a_n = \dfrac{n+1}{n+2} - \dfrac{n}{n+1} = \dfrac{2n-1}{(n+2)(n+1)} > 0$, $a_{n+1} > a_n$, $\{a_n\}$ is an increasing sequence.

- As $|a_n| = \left|\dfrac{n}{n+1}\right| < 1$, the sequence $\{a_n\}$ is a bounded sequence.

2) $a_n = 10 - 2n$

- It is not specified as a finite sequence, then $\{a_n\}$ is an infinite sequence.
- Since $a_{n+1} - a_n = (10 - 2(n+1)) - (10 - 2n) = -2 < 0$, $a_{n+1} < a_n$, $\{a_n\}$ is a decreasing sequence.
- $|a_n| = |10 - 2n|$
  Because there is no positive real number M to let $|10 - 2n| \leqslant M$ when $n \to \infty$, $\{a_n\}$ is not a bounded sequence.
- For $a_{n+1} - a_n = -2$ is a constant, $\{a_n\}$ is an arithmetic sequence.

3) $a_n = \cos n\pi$

- $\{a_n\}$ is an infinite sequence.
- Since $a_{2n} < 0$ and $a_{2n-1} > 0$, $\{a_n\}$ is an oscillating sequence.
- $|a_n| = |\cos n\pi| \leqslant 1$, $\{a_n\}$ is a bounded sequence.

4) $a_n = 3^n$

- Without being specified a finite sequence, $\{a_n\}$ is an infinite sequence.
- $a_{n+1} - a_n = 3^{n+1} - 3^n = 2 \cdot 3^n > 0$,
  Because $a_{n+1} > a_n$, $\{a_n\}$ is an increasing sequence.
- There does not exist a positive real number M to let $|a_n| = |3^n| \leqslant M$ when $n \to \infty$, thus $\{a_n\}$ is not a bounded sequence.
- Because

$$\frac{a_{n+1}}{a_n}=\frac{3^{n+1}}{3^n}=3$$

is a constant, $\{a_n\}$ is a geometric sequence.

5) $a_n=\sin n\pi$ $(1\leqslant n\leqslant 10000)$
   - Since $\{a_n\}$ has 10000 terms it is a finite sequence.
   - For $a_n=\sin n\pi=0$, $\{a_n\}=\{\underbrace{0, 0, 0, \cdots, 0}_{10000}\}$ is a constant sequence.
   - $\{a_n\}$ is a finite sequence, thus it is a bounded sequence.

6) $1, 1^2, 1^3, \cdots, 1^{100}$
   - Since $\{a_n\}$ has 100 terms it is a finite sequence.
   - For $a_{n+1}=a_n$, $\{a_n\}$ is a constant sequence.
   - $\{a_n\}$ is a finite sequence thus it is a bounded sequence.

∎ 1.4 The Sum of the First *n* Terms of a Sequence

There are two types of the sum of an infinite sequences $\{a_n\}$.

- The sum of all terms of an infinite sequence

$$S = a_1 + a_2 + \cdots + a_k + \cdots = \sum_{i=1}^{\infty} a_i.$$

   It is called **series**, which is beyond the scope of this book.

- The sum of the first *n* terms of an infinite sequences or the sum of all terms of a finite sequence consisting such *n* terms.

$$S_n = a_1 + a_2 + \cdots + a_n = \sum_{i=1}^{n} a_i \qquad (1.4.1)$$

   The sum of a sequence discussed in this book falls into this category.

Laws of the Sum of the First *n* Terms of a Sequence

$$\sum_{i=1}^{n} C \cdot a_i = C \sum_{i=1}^{n} a_i \qquad (C \text{ is a constant}) \qquad (1.4.2)$$

$$\sum_{i=1}^{n} (a_i \pm b_i) = \sum_{i=1}^{n} a_i \pm \sum_{i=1}^{n} b_i \qquad (1.4.3)$$

We have two tasks in this section.

1. Find the sum S_n of the first *n* terms of a sequence.
2. Find the general term a_n by the sum S_n.

▶ Find the Sum of the First *n* Terms of a Sequence

We have the following methods to find the sum S_n of a sequences from its general term a_n.

1. Addition by reversing order
2. Elimination by splitting terms
3. Difference by displacing terms

Example 1.4.1   *Addition by Reversing Order*

Find the sum of the first n terms of $\{a_n\}:\{1, 2, 3, \cdots, n, \cdots\}$.

Solution:

$$S_n = 1+2+3+\cdots+(n-1)+n \qquad\qquad (a)$$

Also we can write the above in reverse order

$$S_n = n+(n-1)+\cdots+3+2+1 \qquad\qquad (b)$$

Add (a) and (b).

$$2S_n = (1+n)+(2+(n-1))+(3+(n-2))+\cdots+((n-1)+2)+(n+1)$$
$$= n(n+1)$$

We obtain $\qquad\qquad\qquad S_n = \dfrac{n(n+1)}{2}.$

Example 1.4.2   *Elimination by Splitting Terms*

Find S_n for the following sequences.

1) $a_n = \dfrac{1}{n^2-n}$ $(n>1)$ 2) $a_n = \dfrac{1}{\sqrt{2n}+\sqrt{2n-2}}$

Solution:

1) Split a_n into two parts.

$$\frac{1}{n^2-n} = \frac{1}{n(n-1)}$$
$$= \frac{1}{n-1} - \frac{1}{n}$$

$$S_n = \left(\frac{1}{1}-\frac{1}{2}\right)+\left(\frac{1}{2}-\frac{1}{3}\right)+\left(\frac{1}{3}-\frac{1}{4}\right)+\ldots+\left(\frac{1}{n-1}-\frac{1}{n}\right) = 1-\frac{1}{n}$$
$$= 1-\frac{1}{n}$$

2) Split a_n into two parts.

$$a_n = \frac{1}{\sqrt{2n}+\sqrt{2n-2}} \qquad\qquad (\sqrt{2n}-\sqrt{2n-2}\neq 0)$$
$$= \frac{\sqrt{2n}-\sqrt{2n-2}}{(\sqrt{2n}+\sqrt{2n-2})\cdot(\sqrt{2n}-\sqrt{2n-2})}$$
$$= \sqrt{2n}-\sqrt{2n-2}$$

$$S_n = (\sqrt{2}-0)+(\sqrt{4}-\sqrt{2})+(\sqrt{6}-\sqrt{4})+(\sqrt{8}-\sqrt{6})+\ldots+(\sqrt{2n}-\sqrt{2n-1})$$
$$= \sqrt{2n}.$$

Example 1.4.3   *Difference by Displacing Terms*

If $a_n = 5^n + 2n$ $(n > 0)$, find the sum S_n of the first n terms of the sequence $\{a_n\}$.

Solution:

Let $b_n = 5^n$ and $c_n = 2n$. Then $a_n = b_n + c_n$. Now $\{a_n\}$ is split up into two sequences $\{b_n\}$ and $\{c_n\}$. Let A_n, B_n, and C_n denote the sum of the first n terms of $\{a_n\}$, $\{b_n\}$, and $\{c_n\}$ respectively.

$$A_n = \sum_{i=1}^{n} (5^n + 2n) = \sum_{i=1}^{n} 5^n + 2\sum_{i=1}^{n} n$$
$$= (5^1 + 5^2 + \cdots + 5^n) + 2 \cdot (1 + 2 + 3 + \cdots + n)$$
$$= B_n + C_n$$

We will find B_n and C_n respectively.

- B_n :   Displace each term to next position by multiplying 5 to all terms

  Let
  $$B_n = 5^1 + 5^2 + 5^3 + \cdots + 5^{n-2} + 5^{n-1} + 5^n \tag{a}$$
  $$5B_n = 5^2 + 5^3 + 5^4 + \cdots + 5^{n-1} + 5^n + 5^{n+1} \tag{b}$$

  (b)–(a) :
  $$5B_n - B_n = 5^{n+1} - 5$$
  $$B_n = \frac{5(5^n - 1)}{4}$$

- C_n :   Let $C_n = 1 + 2 + 3 + \ldots + n$, and use the method of reverse addition.
  $$C_n = 1 + 2 + 3 + \cdots + (n-2) + (n-1) + n \tag{c}$$
  $$C_n = n + (n-1) + (n-2) + \cdots + 3 + 2 + 1 \tag{d}$$

  (c)+(d) :
  $$2C_n = n(n+1)$$
  $$C_n = \frac{n(n+1)}{2}$$

We obtain

$$A_n = B_n + 2 \cdot C_n = \frac{5(5^n + 1)}{4} + 2 \cdot \frac{n(n+1)}{2} = \frac{5(5^n - 1)}{4} + n(n+1).$$

▶ Find the General Term from the Sum

The Relationship of a_n and S_n

If S_n is the sum of the first n terms of a sequence $\{a_n\}$, the general term a_n of the sequence $\{a_n\}$ can be expressed as below.

$$a_n = \begin{cases} S_1 & (n=1) \\ S_n - S_{n-1} & (n \geqslant 2) \end{cases} \qquad (1.4.4)$$

The following is a general guide for finding general term a_n of a sequence.
1) Let $n = 1$ and $a_1 = S_1$.
2) Find the sum S_n of the first n terms of the sequence.
3) Express the general term as $a_n = S_n - S_{n-1}$.
4) If a_1 is not suitable to the formula in 3), the piecewise expression (1.4.4) is used.

Example 1.4.4   *Find the General Term from the Sum*

Find general term a_n of a sequence if

1) $S_n = \dfrac{n-1}{2n} \ (n>1)$ 2) $S_n = n^2 + 2n + 1$,

Solution:

1) $a_1 = S_1 = \dfrac{1-1}{2 \cdot 1} = 0$. Since $S_n = \dfrac{n-1}{2n}$ and $S_{n-1} = \dfrac{(n-1)-1}{2(n-1)}$,

$$a_n = S_n - S_{n-1} = \frac{n-1}{2n} - \frac{(n-1)-1}{2(n-1)} = \frac{1}{2n(n-1)}.$$

As a_1 is not suitable to the equation above we obtain the piecewise formula as below.

$$a_n = \begin{cases} 0 & (n=1) \\ \dfrac{1}{2n(n-1)} & (n \geqslant 2) \end{cases}$$

2)
$$a_1 = S_1 = 1^2 + 2 \cdot 1 + 1 = 4$$

Because $S_n = n^2 + 2n + 1$ and $S_{n-1} = (n-1)^2 + 2(n-1) + 1$,
$$a_n = S_n - S_{n-1} = 2n + 1$$

As a_1 is not suitable to the equation above we obtain the piecewise formula as below

$$a_n = \begin{cases} 4 & (n=1) \\ 2n+1 & (n>1) \end{cases}$$

Example 1.4.5   *Find the General Term from the Sum*

If $a_1 = 2$ and $S_n = S_{n-1} + 2a_{n-1} + 2^{n+1}$, find the general term a_n.

Solution:

Using $S_n - S_{n-1} = a_n$, we rewrite $S_n = S_{n-1} + 2a_{n-1} + 2^{n+1}$

to
$$S_n - S_{n-1} = 2a_{n-1} + 2^{n+1}$$
$$a_n - 2a_{n-1} = 2^{n+1}$$

Divide both sides by 2^n.
$$\frac{a_n}{2^n} - \frac{a_{n-1}}{2^{n-1}} = 2$$

Let
$$b_n = \frac{a_n}{2^n}.$$

Since $b_n - b_{n-1} = 2$, then the sequence $\{b_n\}$ is an arithmetic sequence.

For $a_1 = 2$ and $b_1 = \frac{a_1}{2^1} = 1$, $b_n = 1 + (n-1) \cdot 2 = 2n - 1$.

Let
$$\frac{a_n}{2^n} = 2n - 1.$$

Thus
$$a_n = 2^{n+1} n - 2^n.$$

Example 1.4.6   *Find the General Term from the Sum*

If $a_n \geq 1$, $a_1 = 1$, and $S_{n-1} + na_n - 2 = 0$, find the general term a_n.

Solution:

Using $a_n = S_n - S_{n-1}$, we can arrange $S_{n-1} + na_n - 2 = 0$ to
$$n(S_n - S_{n-1}) + S_{n-1} - 2 = 0.$$
$$nS_n - (n-1)S_{n-1} = 2$$

Let
$$b_n = nS_n.$$

Because $b_n - b_{n-1} = 2$, the sequence $\{b_n\}$ is an arithmetic sequence.

For $b_1 = 1 \cdot S_1 = 1 \cdot a_1 = 1$, $b_n = 1 + (n-1) \cdot 2 = 2n - 1$.

Let
$$nS_n = 2n - 1,$$
$$S_n = 2 - \frac{1}{n}.$$

We have
$$a_n = S_n - S_{n-1} = \left(2 - \frac{1}{n}\right) - \left(2 - \frac{1}{n-1}\right) = \frac{1}{(n-1)n}.$$

Therefore
$$a_n = \begin{cases} 1 & (n = 1) \\ \dfrac{1}{(n-1)n} & (n > 1). \end{cases}$$

■ 1.5 Sequences and Functions

By the definition of sequences, a sequence $\{a_n\}$ is a type of discrete functions whose domain is a set of natural numbers.

$$a_n = f(n) \qquad (n \in N^+)$$

Similarly the sum of the first n terms of a sequence is also a discrete function whose domain is a set of natural numbers.

$$S_n = g(n) \qquad (n \in N^+)$$

Because they are limited in their domains of natural number we may face some difficulty in some circumstances such as the following example.

"When f(n) reaches its minimum if $a_n = f(n) = 1 + (n - \frac{15}{2})^2$ $(n \in N^+)$?"

You may say $n = \dfrac{15}{2}$ but it is impossible because n is natural number. In this case, instead of exploring $f(n)$ we will explore a similar function $f(x)$ whose domain is real numbers.

$$f(x) = 1 + (x - \frac{15}{2})^2 \qquad (x \in R)$$

Now we can say "the function $f(x)$ reaches its minimum at $x = \dfrac{15}{2}$". By their similarity we conclude that $a_n = f(n)$ reaches its minimum when n is about $\dfrac{15}{2}$.

As n must be natural number we take 7 and 8 and obtain two minimum terms

$$a_7 = a_8 = 1.25.$$

As $y = f(x)$, called the corresponding function of $a_n = f(n)$, has similar properties as $a_n = f(n)$, we are able to find properties of $a_n = f(n)$ by researching $y = f(x)$. The comparisons between them are given as below.

1) Both $a_n = f(n)$ and $y = f(x)$ have the same function rule f.
2) The function $a_n = f(n)$ is a type of discrete function but $y = f(x)$ is a continuous function.
3) The domain and range of function
   - $a_n = f(n)$
     Domain; $\{n \mid a_n = f(n), n \in N^+\}$
     Range: $\{a_n \mid a_n = f(n), n \in N^+\}$
   - $y = f(x)$

Domain: $\{x \mid y=f(x), x \in R\}$
Range: $\{y \mid y=f(x), x \in R\}$

The domain of $a_n = f(n)$ is a subset of the domain of $y=f(x)$. Similarly the range of $a_n = f(n)$ is a subset of the range of $y=f(x)$. Thus each $a_n = f(n)$ can be find in the range of $y=f(x)$. The graph of $a_n = f(n)$ appears as a series of discrete points on the graph of $y=f(x)$.

▶ Graph of Sequences

Because sequences are functions they can be expressed by graph. The veritable n sits on x-axis and the value of the function, a_n or S_n on y-axis. Both graphs of $\{a_n\}$ and S_n are discrete points. We give some examples as below. The curve is the graph of continuous function, $y=f(x)$, and the dots is the graph of discrete function $a_n = f(n)$.

1) $\{a_n = \dfrac{1}{n}\} : \{1, \dfrac{1}{2}, \dfrac{1}{3}, \cdots, \dfrac{1}{n}, \cdots\}$

From the graph below we know $a_1 = 1$ is the maximum term of $\{a_n\}$ and a_n is approaching 0 when n increases infinitely ($n \to \infty$).

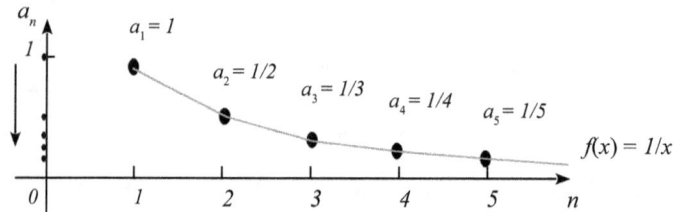

2) $\{a_n = \dfrac{2n-1}{2n+1}\} : \{\dfrac{1}{3}, \dfrac{3}{5}, \dfrac{5}{7}, \dfrac{7}{9}, \dfrac{9}{11}, \cdots, \dfrac{2n-1}{2n+1}, \cdots\}$

From the graph below we know $a_1 = \dfrac{1}{3}$ is the minimum term of $\{a_n\}$ and a_n is approaching 1 when n increases infinitely ($n \to \infty$).

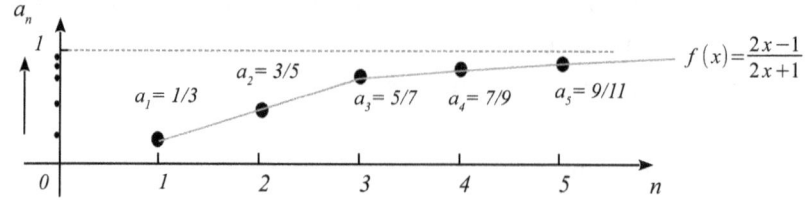

3) $\{a_n = -n^2 + 7n + 1\} : \{7, 11, 13, 13, 11, 7, 1, -7, \cdots, n, \cdots\}$

From the graph below we know $a_3 = a_4 = 13$ are two maximum terms of $\{a_n\}$ and a_n is approaching to the infinity $-\infty$ when n increases infinitely $(n\to\infty)$.

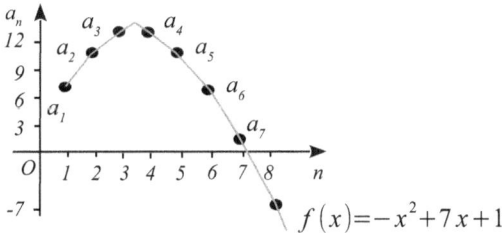

$$f(x) = -x^2 + 7x + 1$$

4) $\{a_n = (-1)^n\} : \{-1, 1, -1, 1, -1, \cdots, (-1)^n, \cdots\}$

When n is increasing, the sequence a_n is oscillate between -1 and 1.

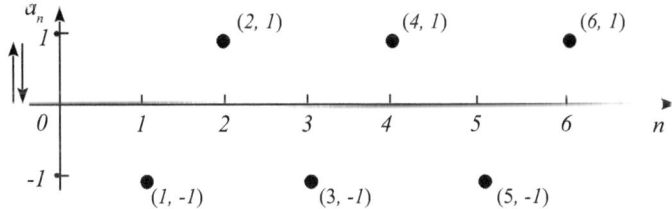

▶ The Extreme of Sequences

As a sequence is a discrete function and we can discuss the extreme (the maximum term or the minimum term) of a sequence like that of functions. First of all we check if a sequence is a monotone sequence, increasing sequence or decreasing sequence.

1. If a sequence $\{a_n\}$ is a monotone sequence we use the following methods.

   - If $a_n < a_{n+1}$ ($\forall n$), the sequence $\{a_n\}$ is an increasing sequence and the first term a_1 is the minimum term in the sequence.

   - If $a_n > a_{n+1}$ ($\forall n$), the sequence $\{a_n\}$ is a decreasing sequence and the first term a_1 is the maximum term in the sequence.

2. If a sequence $\{a_n\}$ is not a monotone sequence we use the following methods. Suppose the term a_n is the maximum (or the minimum) we are looking for in a sequence. Use the following inequalities to find the extreme.

   - To find the maximum term of $\{a_n\}$ let the inequalities below hold.

$$\begin{cases} a_n \geqslant a_{n-1} \\ a_n \geqslant a_{n+1} \end{cases} \qquad (n \geqslant 2) \qquad (1.6.1)$$

- To find the minimum term of $\{a_n\}$ let the inequalities below hold.

$$\begin{cases} a_n \leqslant a_{n-1} \\ a_n \leqslant a_{n+1} \end{cases} \qquad (n \geqslant 2) \qquad (1.6.2)$$

Also we should check if the first term a_l is the extreme term of the sequence since n starts from 2 in the inequalities above.

Example 1.5.1   *Monotonicity of Sequences*

Discuss monotonicity of the sequence $\{a_n = \dfrac{n}{2n+1}\}$ $(n \in N^+)$.

Solution:

We can solve it by two approaches.

1) By the definition of monotone sequences

As
$$a_{n+1} - a_n = \frac{1}{(2n+3)(2n+1)} > 0,$$

$\{a_n\}$ is a monotone increasing sequence.

2) By function method

Let
$$f(x) = \frac{x}{2x+1}$$

be the corresponding function of

$$a_n = f(n) = \frac{n}{2n+1}$$

and we can discuss monotonicity of the function

$$f(x) = \frac{x}{2x+1} \quad (x > 0).$$

As
$$f(x) = \frac{x}{2x+1} = \frac{\frac{1}{2}(2x+1) - \frac{1}{2}}{2x+1} = \frac{1}{2} - \frac{\frac{1}{2}}{2x+1},$$

The function $f(x)$ is a monotone increasing function on domain $(0, +\infty)$. Thus its corresponding function $a_n = f(n)$ is a monotone increasing function too. Then the sequence $\{a_n\}$ is a monotone increasing sequence.

Example 1.5.2   *The Extreme of Sequences*

1) If $a_n = 3n^2 - 2n + 1$, find the extreme term of $\{a_n\}$.

2) If $a_n = -4n^2 + 5n + 2$, find the extreme term of $\{a_n\}$.

Solution:

1) Check if the sequence $\{a_n\}$ is a monotone sequence.

$$\begin{cases} a_n = 3n^2 - 2n + 1 \\ a_{n-1} = 3(n-1)^2 - 2(n-1) + 1 \end{cases}$$

$$a_n - a_{n-1} = 6n - 5 > 0 \quad (n \geqslant 2)$$

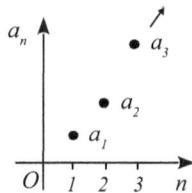

The sequence $\{a_n\}$ is an increasing sequence and the first term $a_1 = 2$ is the minimum term.

2) Check if the sequence $\{a_n\}$ is a monotone sequence.

$$\begin{cases} a_n = -4n^2 + 5n + 2 \\ a_{n-1} = -4(n-1)^2 + 5(n-1) + 2 \end{cases}$$

$$a_n - a_{n-1} = 9 - 8n < 0 \quad (n \geqslant 2)$$

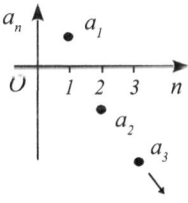

The sequence $\{a_n\}$ is a decreasing sequence and the first term $a_1 = 3$ is the maximum term in the sequence.

Example 1.5.3   *The Extreme of Sequences*

1) If $a_n - \lambda n^2 - 5n + 1$ and the term a_{10} is the minimum in the sequence, find the range of real number λ.

2) If $a_n = 1 + (n - \dfrac{17}{2})^2$, find the extreme term of the sequence.

Solution:

1) Because the term a_{10} is the minimum in the sequence it should be smaller than its two neighbors. Let the following inequalities hold.

$$\begin{cases} a_{10} \leqslant a_9 \\ a_{10} \leqslant a_{11} \end{cases}$$

$$\begin{cases} 10^2 \lambda \quad 49 \leqslant 9^2 \lambda - 44 \\ 10^2 \cdot \lambda - 49 \leqslant 11^2 \lambda - 54 \end{cases}$$

We obtain
$$\frac{5}{21} \leqslant \lambda \leqslant \frac{5}{19}.$$

If the value of λ falls in the range above the term a_{10} is the minimum in the sequence.

2) Obviously the sequence $\{a_n\}$ has its minimum when $n=\dfrac{17}{2}$ but n is a natural number. We can use a like function

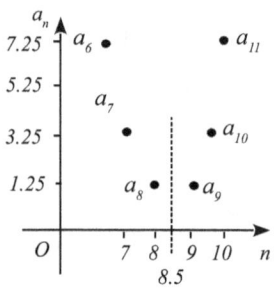

$$y=1+\left(x-\dfrac{17}{2}\right)^2 \quad (x \in R).$$

We know that y reaches its minimum, $y=1$, when $x=8.5$. Accordingly the sequence $\{a_n\}$ has its minimum(s) when n is about 8.5. As n is a natural number, we take 8 and 9. We obtain the minimum terms $a_8=a_9=1.25$ in the sequence.

Example 1.5.4   *The Extreme of Sequences*

If $a_n=-n^2+7n+1$, find the extreme term of the sequence.

Solution:

Check if $\{a_n\}$ is a monotone sequence.

As
$$\begin{cases} a_n=-n^2+7n+1 \\ a_{n-1}=-(n-1)^2+7(n-1)+1 \end{cases}$$

$$a_n-a_{n-1}=8-2n \quad (n \geqslant 2).$$

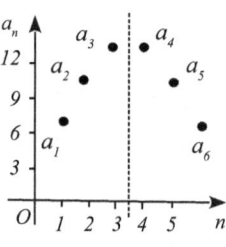

When $2 \leqslant n < 4$, $a_n-a_{n-1}>0$ and $\{a_n\}$ increases. When $n>4$, $a_n-a_{n-1}<0$ and $\{a_n\}$ decreases. Thus $\{a_n\}$ is not a monotone sequence and has its maximum. Since the first term a_1 is not the extreme we suppose the term a_n is the maximum in the sequence.

Let
$$\begin{cases} a_n \geqslant a_{n-1} & (n \geqslant 2) \\ a_n \geqslant a_{n+1} \end{cases}$$

$$\begin{cases} -n^2+7n+1 \geqslant -(n-1)^2+7(n-1)+1 \\ -n^2+7n+1 \geqslant -(n+1)^2+7(n+1)+1 \end{cases}$$

We obtain $3 \leqslant n \leqslant 4$ and take $n = 3$ and 4. Therefor the 3rd term $a_3=13$ and the 4th term $a_4=13$ are the maximum terms in the sequence.

Example 1.5.5   *The Extreme of Sequences*

If $a_n=\left(n-\dfrac{5}{2}\right)^2+2$, find the extreme term of the sequence.

Solution:

Check if the sequence $\{a_n\}$ is a monotone sequence.

As
$$\begin{cases} a_n=\left(n-\dfrac{5}{2}\right)^2+2 \\ a_{n-1}=\left((n-1)-\dfrac{5}{2}\right)^2+2 \end{cases}$$

$$a_n-a_{n-1}=2n-6 \quad (n\geqslant2).$$

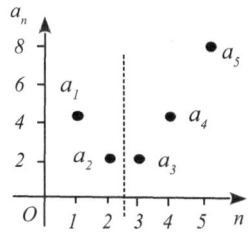

When $2\leqslant n<3$, $a_n-a_{n-1}<0$ and the sequence $\{a_n\}$ decreases. When $n>3$, $a_n-a_{n-1}>0$ and the sequence $\{a_n\}$ increases (See the figure below). Thus the sequence $\{a_n\}$ is not a monotone sequence and has its minimum. Since the first term a_1 is not the extreme we suppose the term a_n is the minimum.

Let
$$\begin{cases} a_n\leqslant a_{n-1} \\ a_n\leqslant a_{n+1} \end{cases} \quad (n\geqslant2)$$

$$\begin{cases} \left(n-\dfrac{5}{2}\right)^2+2\leqslant\left((n-1)-\dfrac{5}{2}\right)^2+2 \\ \left(n-\dfrac{5}{2}\right)^2+2\leqslant\left((n+1)-\dfrac{5}{2}\right)^2+2 \end{cases}$$

We obtain $2\leqslant n\leqslant3$. As n is a nature number we take 2 and 3. Then $a_2=2\dfrac{1}{4}$ and $a_3=2\dfrac{1}{4}$ are the minimum terms in the sequence.

∎ 1.6 Term Groups

We introduce a new terminology, the **term group** of a sequence. This concept will help us to explore more properties of sequences. The content of this part is an extra for some readers of high level.

DEFINITION   Term Group

A term group of a sequence is a group of m ($m \in N^+$, $1 < m < n$) consecutive terms from the sequence, which keep their original order in the group unchanged.

You can construct term groups within a sequence by whatever rule you set. But we are interested in a specific type of term group used later thus we set the following rules.

Rules of Constructing Term Groups

1) All term groups have the same size, m ($1 < m < n$) consecutive terms.

2) There may have an interval of p ($0 \leq p \leq n-2$) terms between two adjacent term groups.

3) The first term group is not necessary to start from the first term of $\{a_n\}$.

If the kth term group of the sequence $\{a_n\}$ is denoted by

$$\{g_k\} : \{a_n, a_{n+1}, \cdots, a_{n+m-1}, a_{n+m-1}\} \qquad (k \in N^+, 1 < m < n),$$

we can write out all other term groups like

$$\{g_{k-1}\} : \{a_{n-p-m}, a_{n-p-m+1}, \cdots, a_{n-p-2}, a_{n-p-1}\} \quad (p \geq 0)$$
$$\{g_{k+1}\} : \{a_{n+p+m}, a_{n+p+m+1}, \cdots, a_{n+p+2m-2}, a_{n+p+2m-1}\}.$$

The figure below will help you to understand the concept.

Now we obtain a new sequence $\{G_k\}$ containing all term groups.

$$\{G_k\}:\{\ \{g_1\},\ \{g_2\},\ \cdots,\ \{g_{k-1}\},\ \{g_k\},\ \{g_{k+1}\},\ \cdots\}.$$

Let's take a example. Suppose the sequence $\{a_n\} = \{1, 2, 3, 4, 5, 6,\ \cdots\}$. Then the term groups of the sequence $\{a_n\}$ can be constructed like one of the following patterns.

- $m = 5$ and $p = 0$:

$$\{G_k\}:\{\ \underbrace{\{1, 2, 3, 4, 5\}}_{\{g_1\}},\ \underbrace{\{6, 7, 8, 9, 10\}}_{\{g_2\}},\ \underbrace{\{11, 12, 13, 14, 15\}}_{\{g_3\}},\ \cdots\ \}.$$

- $m = 3$ and $p = 5$:

$$\{G_k\}:\{\ \underbrace{\{4, 5, 6\}}_{\{g_1\}},\ \underbrace{\{12, 13, 14\}}_{\{g_2\}},\ \underbrace{\{20, 21, 22\}}_{\{g_3\}},\ \underbrace{\{28, 29, 30\}}_{\{g_3\}},\ \cdots\ \}.$$

2 Arithmetic Sequences

▌2.1 Definition

Let's look at some examples of arithmetic sequence.

1) $1, 2, 3, 4, 5, \cdots, n, \cdots$ $(2-1 = 3-2 = \cdots = n - (n-1) = 1)$

2) $3, 6, 9, 12, 15, \cdots, 3n, \cdots$ $(6-3 = 9-6 = \cdots = 3n - 3(n-1) = 3)$

3) $3, 1, -1, -3, -5, \cdots, 5 - 2n, \cdots$ $(1-3 = -1-1 = -3 - (-1) = \cdots = -2)$

These sequences have a common that starting from the second term, the differences between any two consecutive terms are the same.

DEFINITION   Arithmetic Sequences

A sequence $\{a_n\}:\{a_1, a_2, \cdots, a_n, \cdots\}$ is an arithmetic sequence if

$$a_{n+1} - a_n = d \qquad (n \in N^+) \qquad (2.1.1)$$

Where d is a constant and called **common difference**.

Example 2.1.1   *Definition of Arithmetic Sequences*

Find which sequence given below is an arithmetic sequence?

  1) $a_n = 3n$ 2) $a_n = 5 - 2n$ 3) $a_n = 2n^2 + 3$

Solution:

1) $\{a_n\}: \{ 3,\quad 6,\quad 9,\quad 12,\quad \cdots, 3n, 3(n+1), \cdots \}$

  $6-3=3 \quad 9-6=3 \quad 12-9=3 \qquad 3(n+1)-3n=3$

Because $d = a_{n+1} - a_n = 3$ is a constant, $\{a_n\}$ is an arithmetic sequence.

2) $\{a_n\}: \{ 3,\quad 1,\quad -1,\quad -3, \cdots, 5-2n, 5-2(n+1), \cdots \}$

  $1-3=-2 \quad -1-1=-2 \quad -3-(-1)=-2 \qquad 5-2(n+1)-(5-2n) = -2$

Since $d = a_{n+1} - a_n = -2$ is a constant, $\{a_n\}$ is an arithmetic sequence.

3) $\{a_n\}: \{ 5,\quad 11, 21, \cdots, 2n^2 + 3, \cdots \}$

$$d = a_{n+1} - a_n = 2(n+1)^2 + 3 - (2n^2 + 3) = 4n + 2.$$

Because d is not a constant, $\{a_n\}$ is not an arithmetic sequence.

■ 2.2 General Term

Let's figure out the formula of general term (the nth term) of an arithmetic sequence. By the definition of arithmetic sequences, we list all terms of an arithmetic sequence as below.

$$a_1 = a$$
$$a_2 = a_1 + d$$
$$a_3 = a_2 + d = a_1 + 2d$$
$$\cdots$$
$$a_n = a_{n-1} + d = a_1 + (n-1)d$$
$$\cdots$$

Now we find that the formula of the general term a_n (the nth term) is related to three basic facts a_1, n, and d with which an arithmetic sequence can be determined.

The General Term of Arithmetic Sequences

The formula of the general term a_n of an arithmetic sequence $\{a_n\}$ is a linear function in terms of n. If a_1 is the first term and d is the common difference, the general term of $\{a_n\}$ is

$$a_n = a_1 + (n-1)d \qquad (n \in N^+) \qquad (2.2.1)$$

When the subscription of the first term starts at m, the general term of $\{a_n\}$ is

$$a_n = a_m + (n-m)d \qquad (m, n \in N^+, \ m < n) \quad (2.2.2)$$

Arrange the formula (2.2.1) to

$$a_n = a_1 + (n-1)d = d \cdot n + (a_1 - d).$$

Let $p = d$ and $q = a_1 - d$, we have the following conclusion.

THEOREM 2.2.1

The formula of general term of an arithmetic sequence $\{a_n\}$ is a linear function in terms of n.

$$a_n = p \cdot n + q \quad (p, q \text{ are constants}, \ p, q \neq 0) \qquad (2.2.3)$$

Example 2.2.1   *General Term of Arithmetic Sequences*

If the sequence $\{a_n\}$ is an arithmetic sequence,

    1) $a_1 = 3, d = -6$, find a_8.
    2) $a_2 = 12, d = 4$, find a_{10}.
    3) $a_3 = 10, a_9 = 28$, find a_{24}.
    4) $a_1 = 4, d = 2$, which term is great than 40?

Solution:

1) As $a_n = 3 + (n-1)(-6)$, $a_8 = 3 + (8-1)(-6) = 39$.

2) As $a_2 = a_1 + d$ and $a_1 = a_2 - d = 8$, $a_n = 8 + (n-1)\cdot 4$.

$$a_{10} = 8 + (10-1)\cdot 4 = 44$$

3) Since $a_9 - a_3 = 6d = 18$, $d = 3$ and $a_1 = 4$, $a_n = 4 + (n-1)\cdot 3$.

$$a_{24} = 4 + 23\cdot 3 = 73$$

4) Let $a_n = 4 + (n-1)\cdot 2 > 40$, we obtain $n = 20$.

Example 2.2.2   *General Term of Arithmetic Sequences*

Suppose $\{a_n\}$ is an arithmetic sequence.

    1) $a_1 = 3$ and $d = 5$, find a_n.
    2) the first term $a_{100} = 220$ and $d = -2$, find a_n $(n \geqslant 100)$.
    3) $a_7 = 86$ and $a_{17} = 26$, find a_{31}.
    4) $a_n = 2 + 3n$ $(n \geqslant 1)$, find a_1 and d.
    5) $a_n = 8 + (n+1)\cdot 2$ $(n \geqslant 10)$, find a_{10} and d.
    6) $a_2 \cdot a_4 = 96$, $a_5 + a_9 = 36$, find a_7.

Solution:

1) By the formula (2.2.1), $a_n = 3 + (n-1)\cdot 5 = 5n - 2$.

2) Since $a_{100} = 220$ is the first term, we use the formula (1.2.3) of general term.

$$a_n = 220 + (n-100)\cdot (-2) = 420 - 2n \qquad (n \geqslant 100)$$

3) By the formula (2.2.1), $a_{17} - a_7 = 10\cdot d = 26 - 86 = -60$. We have $d = -6$ and

$$a_{31} - a_{17} = 14\cdot d = 14\cdot(-6) = -84$$
$$a_{31} = a_{17} - 84 = 26 - 84 = -58.$$

4) $a_n = 2 + 3n = 5 + (n-1)\cdot 3$

Compare it with the formula (2.2.1), we obtain $a_1 = 5$ and $d = 3$.

5) To match the formula (2.2.1) we rewrite $a_n = 8 + (n+1)\cdot 2$ as

$$a_n = 10 + 2n = 10 + 20 + 2n - 20 = 30 + (n-10)\cdot 2$$

Then $a_{10} = 30$ and $d = 2$.

6) As $a_2=a_1+d$, $a_4=a_1+3d$, $a_5=a_1+4d$, $a_9=a_1+8d$, we have two equations.

$$\begin{cases} a_2 \cdot a_4 = (a_1+d)(a_1+3d) = 96 & \text{(a)} \\ a_5+a_9 = a_1+4d+a_1+8d = 36 & \text{(b)} \end{cases}$$

From (b), $a_1 = 18-6d$ and substitute it for a_1 of (a). Then we obtain a quadratic equation

$$15\,d^2 - 144\,d + 228 = 0.$$

There are two real roots 7.6 and 2. Take $d = 2$ and substitute it for d of (b). We obtain $a_1 = 6$ and

$$a_n = 6 + (n-1)\cdot 2.$$

Then
$$a_7 = 6 + (7-1)\cdot 2 = 18.$$

Example 2.2.3   *General Term of Arithmetic Sequences*

If $\{a_n\}:\{6,\ 8,\ 10,\ 12,\ ...\}$ and $\{b_n\}:\{15,\ 18,\ 21,\ 24,\ ...\}$ are arithmetic sequences and both have 10 terms, find how many terms of both sequences are in common.

Solution:

Because $a_n = 6 + (n-1)\cdot 2$ and $b_m = 15 + (m-1)\cdot 3$, let $a_n = b_m$.

$$2n+4 = 3m+12$$

Then
$$n = 3\frac{m}{2} + 4.$$

Since $1 \leqslant n \leqslant 10$, let
$$1 \leqslant 3\frac{m}{2} + 4 \leqslant 10.$$

Because n and m should be natural numbers we have $m = 2$ and 4. Then there are two terms from two sequences are in common, $a_7 = b_2 = 18$ and $a_{10} = b_4 = 24$.

Example 2.2.4   *General Term of Arithmetic Sequences*

Suppose that the sequence $\{a_n\}$ is an arithmetic sequence, $a_{11}=36$, and $a_{35}=84$.
1) Find a_{2017} 2) Determine which term is *212* 3) Is   *599*   a   term   of   the sequence?

Solution:

Since the sequence $\{a_n\}$ is an arithmetic sequence,

$$\begin{cases} a_{11} = a_1 + (11-1)d = 36 \\ a_{35} = a_1 + (35-1)d = 84 \end{cases}$$

Then $a_1 = 16$ and $d = 2$, we have $a_n = 16 + (n-1)\cdot 2$.   (a)

1)
$$a_{2017} = 16 + (2017-1)\cdot 2 = 4048.$$

2) Because the general term is $a_n = 16 + (n-1)\cdot 2$, substitute 212 for a_n of (a).

$$212=16+(n-1)\cdot 2$$

Then $n = 99$, $a_{99}=212$.

3) Substitute 599 for a_n of (a), $599=16+(n-1)\cdot 2$. We obtain $n = 292.5$. Since n should be a natural number, 599 is not a term of the sequence.

Example 2.2.5   General Term of Arithmetic Sequences

Suppose the sequence $\{a_n\}$ is an arithmetic sequence and $a_1=\dfrac{1}{10}$. If $a_n > 1$ ($\forall n \geqslant 11$), what is the range of its common difference d?

Solution:

If $n \geqslant 11$, let

$$a_n=\frac{1}{10}+(11-1)d>1$$

and

$$d>\frac{9}{100}$$

else let

$$a_n=\frac{1}{10}+(10-1)d\leqslant 1$$

and

$$d\leqslant\frac{1}{10}.$$

Then we have the range of the common difference, $\dfrac{9}{100}<d\leqslant\dfrac{1}{10}.$

Example 2.2.6   General Term of Arithmetic Sequences

If the sequence $\{2a_1+1,2a_2+2,2a_3+3,...,2a_n+n,...\}$ is an arithmetic sequence with common differences $d = 5$ and $a_1 = 1$, find
  1) the 10th term of the sequence $\{2a_n+n\}$ 2) a_n.

Solution:

The sequence $\{2a_n+n\}$ is an arithmetic sequence with the first term $2a_1+1=3$ and common difference $d=5$. Then the general term is $2a_n+n$ and

$$2a_n+n=3+(n-1)\cdot 5.$$

1) The 10th term of $\{2a_n+n\}$ becomes

$$2a_{10}+10 = 3+(10-1)\cdot 5 = 48.$$

2) By the definition, we have $(2a_n+n)-(2a_{n-1}+(n-1))=5$.

$$a_n-a_{n-1}=2.$$

Then the sequence $\{a_n\}$ is an arithmetic sequence whose $a_1=1$ and common difference $d=2$. Thus

$$a_n=1+(n-1)2=2n-1.$$

2.3 The Sum of the First *n* Terms

Now we discuss the sum of the first *n* terms of an infinite arithmetic sequences. Suppose the sequence $\{a_n\}$ is an infinite arithmetic sequence. By equation (1.1.3), we have the sum of the first *n* terms of the sequence $\{a_n\}$ as below.

$$S_n = a_1 + a_2 + a_3 + \cdots + a_n \qquad \text{(a)}$$

On other hand, reverse the order of S_n, then we have

$$S_n = a_n + a_{n-1} + a_{n-2} + \cdots + a_1 \qquad \text{(b)}$$

Add (a) and (b), $2S_n = (a_1 + a_n) + (a_2 + a_{n-1}) + \cdots + (a_n + a_1) = n(a_1 + a_n)$

$$S_n - \frac{n \cdot (a_1 + a_n)}{2}$$

The Sum of the First *n* Terms of an Arithmetic Sequence

The sum of the first *n* terms of an arithmetic sequence $\{a_n\}$ is

$$S_n = \frac{n(a_1 + a_n)}{2} \qquad (n \in N^+) \qquad (2.3.1)$$

or

$$S_n = n a_1 + \frac{n(n-1)d}{2} \qquad (2.3.2)$$

The formula of the sum S_n is a quadratic function in terms of *n*.

The Sum of *m* Consecutive Terms of an Arithmetic Sequence

Let S_m be the sum of *m* consecutive terms starting at term a_k of an arithmetic sequence $\{a_n\}$, then

$$S_m = \frac{m(a_k + a_{k+m-1})}{2}. \qquad (2.3.3)$$

or

$$S_m = m a_k + \frac{m(m-1)d}{2}. \qquad (1 \leqslant k < n-m-1) \quad (2.3.4)$$

Look at the formulas above there are five basic quantities, a_1, a_n, *n* (or *m*), *d*, and S_n. We can obtain any two from other three. We rearrange (2.3.2) to

$$S_n = \frac{d}{2} n^2 + \left(a_1 - \frac{d}{2}\right) n.$$

Let $p = \dfrac{d}{2}$ and $q = a_1 - \dfrac{d}{2}$, then we have the following conclusion.

> **THEOREM 2.3.1**
>
> The sum of the first n terms of the sequence $\{a_n\}$ is a quadratic function in terms of n without constant term.
>
> $$S_n = pn^2 + qn \qquad (\forall n) \qquad\qquad (2.3.5)$$

Example 2.3.1   *The Sum of the First n Terms*

The sequence $\{a_n\}$ is an arithmetic sequence, $a_1 = 6$, and $a_{11} = 36$, find

  1) $S_{20} = ?$ 2) $a_{101} + a_{102} + a_{103} + a_{104} + a_{105} = ?$

Solution:

1) According to the formula (2.2.1), $a_{11} = 6 + (11-1)d = 36$ and $d = 3$.

By the formula (2.3.2), $S_n = 6n + \dfrac{n(n-1)\cdot 3}{2}$.

Then $S_{20} = 690$.

2) Let $\qquad\qquad\qquad S = a_{101} + a_{102} + a_{103} + a_{104} + a_{105}.$

Since $a_{101} = 6 + (101-1)\cdot 3 = 306$ and $m = 5$, by the formula (2.3.4),

$$S = 5(306) + \frac{5(5-1)\cdot 3}{2} = 1560.$$

Example 2.3.2   *The Sum of the First n Terms*

Suppose the sequence $\{a_n\}$ is an arithmetic sequence.

  1) $\{a_n\} = 3, 8, 13, \cdots, a_n = a_{n-1} + 5, \cdots$, find S_n.
  2) $S_n = 2n^2$, find a_n.
  3) $a_1 + a_2 + a_3 = 6$, $a_{n-2} + a_{n-1} + a_n = 30$, and $S_n = 120$, $n = ?$
  4) $a_3 + a_7 = 12$, $a_8 + a_{12} = 72$, $S_{10} = ?$
  5) $S_4 = 16$, $S_{12} = 144$, find $a_{101} + a_{102} + a_{103} + a_{104} = ?$
  6) $a_1 + a_4 + a_7 + a_{10} = 16$, find S_{10}.

Solution:

1) Since $a_1 = 3$, $d = a_n - a_{n-1} = 5$,

$$S_n = \frac{n(2a_1 + (n-1)d)}{2} = \frac{n(5n-1)}{2}.$$

2) Because $a_1 = S_1 = 2\cdot 1^2 = 2$, $S_n = \dfrac{n(a_1 + a_n)}{2} = \dfrac{n(2+a_n)}{2}$.

Let $\qquad\qquad\qquad \dfrac{n(2+a_n)}{2} = 2n^2$

and we have $\qquad\qquad a_n = 4n - 2.$

3) By the definition, $a_1 + a_n = a_2 + a_{n-1} = a_3 + a_{n-2}$.

Because $a_1 + a_n = \dfrac{6+30}{3} = 12$, $S_n = \dfrac{n(a_1 + a_n)}{2} = 6n = 120$. We obtain $n = 20$.

4) Let
$$\begin{cases} a_3 + a_7 = (a_1 + (3-1)d) + (a_1 + (7-1)d) = 12 \\ a_8 + a_{12} = (a_1 + (8-1)d) + (a_1 + (12-1)d) = 72 \end{cases}$$

Since $a_1 = -18$ and $d = 6$, $S_{10} = 90$.

5) As $S_4 = 4a_1 + \dfrac{4 \cdot 3}{2}d = 16$ and $S_{12} = 12a_1 + \dfrac{12 \cdot 11}{2}d = 144$, we obtain $a_1 = 1$ and

$d = 2$. By the formula (2.2.1), $a_{101} = 201$ and $a_{104} = 207$. By the formula (2.3.1),

$$a_{101} + a_{102} + a_{103} + a_{104} = \dfrac{4 \cdot (a_{101} + a_{104})}{2} = 816.$$

6) Because
$$\begin{cases} a_1 + a_{10} = a_1 + a_1 + (10-1)d = 2a_1 + 9d \\ a_4 + a_7 = a1 + (4-1)d + a_1 + (7-1)d = 2a_1 + 9d, \end{cases}$$

$$a_1 + a_{10} = a_4 + a_7 = 16/2 = 8.$$

Then
$$S_{10} = \dfrac{10(a_1 + a_{10})}{2} = \dfrac{10(8)}{2} = 40.$$

Example 2.3.3   *The Sum of the First n Terms*

Suppose that S_n is the sum of the first n terms of the arithmetic sequence $\{a_n\}$, $a_3 = 7$, $S_7 = 63$, and $S_k = 35$. Find k.

Solution:

By the general term a_n, $a_3 = a_1 + 2d = 7$ (a)

By the formula (2.3.3) $S_7 = 7a_1 + \dfrac{7 \cdot 6 \cdot d}{2} = 63$.

$$S_7 = a_1 + 3d = 9.$$ (b)

Solve the equation set of (a) and (b)
$$\begin{cases} a_1 + 2d = 7 \\ a_1 + 3d = 9 \end{cases}$$

As $a_1 = 3$ and $d - 2$, $S_n = n^2 + 2n$. Let $S_k = 35$, then we have a quadratic equation:
$$k^2 + 2k - 35 = 0.$$

Let the discriminant of this quadratic equation
$$\Delta = b^2 - 4ac = 2^2 - 4 \cdot (-35) = 144 > 0.$$

We obtain two solutions $k_{1,2} = \dfrac{-2 \pm 12}{2}$ and take $k = 5$.

Example 2.3.4   *The Sum of the First n Terms*

Suppose that S_n is the sum of the first n terms of a sequence $\{a_n\}$ and T_n is the sum of the first n terms of the sequence $\{S_n\}$. If $a_1 = 2$, $\{S_n\}$ is an arithmetic sequence, and $T_9 = 90$, find the general term a_n.

Solution:

We can not assume that the sequence $\{a_n\}$ is an arithmetic sequence. As the sequence $\{S_n\}$ is an arithmetic sequence and $S_1 = a_1 = 2$,

$$S_n = S_1 + (n-1) \cdot d_s = 2 + (n-1) \cdot d_s$$

Since $T_n = n S_1 + \dfrac{n(n-1) \cdot d_s}{2}$, $T_9 = 9 \cdot 2 + \dfrac{9 \cdot 8 \cdot d_s}{2} = 90$.

We obtain the common difference $d_s = 2$ for the sequence $\{S_n\}$.

Then
$$\begin{cases} S_n = 2 + (n-1) \cdot 2 = 2n \\ S_{n-1} = 2(n-1) = 2n - 2 \end{cases}$$

$$a_n = S_n - S_{n-1} = 2 \qquad\qquad (n \geqslant 2)$$

Therefore the sequence $\{a_n\}$ is a constant sequence $\{2, 2, 2, 2, \cdots\}$.

Example 2.3.5   *The Sum of the First n Terms*

Let S_n be the sum of the first n terms of an arithmetician sequence $\{a_n\}$. $S_9 = 72$, find a_5.

Solution:

By the formula (2.3.3), let $S_9 = 9 a_1 + \dfrac{9(9-1) \cdot d}{2} = 9(a_1 + 4 \cdot d) = 72$.

$$a_1 + 4 \cdot d = 8 \qquad\qquad\qquad\qquad (a)$$

On the other hand, $\qquad a_5 = a_1 + (5-1) \cdot d = a_1 + 4 \cdot d$. $\qquad\qquad$ (b)

Compare (a) with (b), we obtain $a_5 = 8$.

Example 2.3.6   *The Sum of the First n Terms*

Suppose that S_n is the sum of the first n terms of an arithmetic sequence $\{a_n\}$,

$S_n = \left(\dfrac{a_n + 1}{2}\right)^2$ $(n \in N^+)$, and $b_n = \dfrac{3 S_n}{n} + 4$. Find the sum B_n of the first n terms of the sequence $\{b_n\}$.

Solution:

As $S_n = \left(\dfrac{a_n + 1}{2}\right)^2$ and $S_{n+1} = \left(\dfrac{a_{n+1} + 1}{2}\right)^2$, $a_{n+1} = S_{n+1} - S_n = \left(\dfrac{a_{n+1} + 1}{2}\right)^2 - \left(\dfrac{a_n + 1}{2}\right)^2$.

Then $\qquad\qquad 4 a_{n+1} = a_{n+1}^2 - a_n^2 + 2 a_{n+1} - 2 a_n + a_{n+1} a_n - a_{n+1} a_n$

2.3 The Sum of the First n Terms

$$(a_{n+1}+a_n)(a_{n+1}-a_n-2)=0$$

We have $a_{n+1}+a_n=0$ or $a_{n+1}-a_n-2=0$. Because the sequence $\{a_n\}$ is an arithmetic sequence we take $a_{n+1}-a_n-2=0$. Then $d=a_{n+1}-a_n=2$.

As
$$a_1=S_1=(\frac{a_1+1}{2})^2,$$

we have $a_1^2-2a_1+1=0$.

$$(a_1-1)^2=0.$$

Thus $a_1=1$ and we have
$$a_n=1+(n-1)\cdot2$$
$$a_n=2n-1.$$

S_n becomes
$$S_n=\frac{n(a_1+a_n)}{2}=n^2.$$

Substitute it for S_n in the formula
$$b_n=\frac{3S_n}{n}+1-3n+1.$$

As
$$b_{n+1}=3(n+1)+4=3n+7,$$
$$b_{n+1}-b_n=3=d.$$

By the definition of arithmetic sequences, the sequence $\{b_n\}$ is an arithmetic sequence and $b_1=3S_1+4=3+4=7$. Therefore the sum of the first n terms of the sequence $\{b_n\}$ is
$$B_n=n+\frac{n(n-1)\cdot3}{2}=\frac{3}{2}n^2-\frac{1}{2}n.$$

Example 2.3.7   The Sum of the First n Terms

Let S_n be the sum of the first n terms of an arithmetician sequence $\{a_n\}$. If $S_9=198$ and $S_{18}=639$, find S_{99}.

Solution:

By the formula (2.3.2), we have
$$\begin{cases} S_9=9a_1+\dfrac{9\cdot8\cdot d}{2}=198 \\ S_{18}=18a_1+\dfrac{18\cdot17\cdot d}{2}=639. \end{cases}$$

We have $a_1=10$ and $d=2$. By the formula (2.3.5)
$$S_n=\frac{3}{2}n^2+(10-\frac{3}{2})n.$$

$$S_{99}=15543.$$

▋ 2.4 Arithmetic Mean

At first we give the definition of arithmetic mean of two numbers then we extend it to two terms of arithmetic sequences.

DEFINITION   Arithmetic Mean of Two Numbers

For any three numbers a, b, and c, if the number b is the average of a and c, the number b is called the **arithmetic mean** of a and c.

$$b=\frac{a+c}{2}$$ (2.4.1)

Example 2.4.1   *The Arithmetic Mean of Two Numbers*

Find the arithmetic mean of following two numbers.

1) 3 and 9 2) 2^2 and 2^6 3) $a + b$ 4) $\sqrt{2}$ and $\sqrt{3}$

Solution:

1) $\dfrac{3+9}{2}=6$ 2) $\dfrac{2^2+2^6}{2}=17$ 3) $\dfrac{a+b}{2}$ 4) $\dfrac{\sqrt{2}+\sqrt{3}}{2}$

The same relationship exists in an arithmetic sequence. Let the term a_n be the mid-term, the terms a_{n-m} and a_{n+m} $(m\geqslant1)$ are symmetric about the term a_n. In other words, the term a_{n-m} and a_{n+m} have the same distance to the term a_n. Then the term a_n is the arithmetic mean of a_{n-m} and a_{n+m}.

$$a_1, a_2, a_3, \cdots ,\boxed{a_{n-m}}, \cdots a_{n-1},\boxed{a_n,}a_{n+1}, \cdots ,\boxed{a_{n+m},} \cdots$$

a m terms b m terms c

Because

$$\begin{cases} a_{n-m}=a_1+(n-m-1)d \\ a_{n+m}=a_1+(n+m-1)d \end{cases}$$

$$a_{n-m}+a_{n+m}=a_1+(n-m-1)d+a_1+(n+m-1)d$$
$$=2(a_1+(n-1)d)=2a_n.$$

Then

$$a_n=\frac{a_{n-m}+a_{n+m}}{2}.$$

Arithmetic Mean of Two Terms of Arithmetic Sequences

For three terms a_{n-m}, a_n, and a_{n+m} of an arithmetic sequence $\{a_n\}$ the following equation holds.

$$a_n = \frac{a_{n-m} + a_{n+m}}{2} \quad (\forall n, n \in N^+, m \geqslant 1) \quad (2.4.2)$$

The term a_n is called the **arithmetic mean** of the terms a_{n-m} and a_{n+m}.

Example 2.4.2   Arithmetic Mean of Two Terms of A. S.

If $\{a_n\}$ is an arithmetic sequence, determine if the following statements are true?

1) $2a_6 = a_2 + a_{10}$ 2) $2a_6 = a_2 + a_8$ 3) $2a_n = a_{n-5} + a_{n+5}$ 4) $2a_n = a_{n-m} + a_{n+m+1}$

Solution:

If both terms a_{n-m} and a_{n+m} are symmetric about the term a_n, the statement is true.

Answer	Three subscripts	The symmetry of subscripts
1) True.	$2, 6, 10,$	$6-2=10-6$
2) False.	$2, 6, 8$	$6-2 \neq 8-6$
3) True.	$n-5,\ n,\ n+5$	$n-(n-5)=(n+5)-n$
4) False.	$n-m,\ n,\ n+m+1$	$n-(n-m) \neq (n+m+1)-n$

Example 2.4.3   Arithmetic Mean of Two Terms of A. S.

1) If three numbers a, b, c form an arithmetic sequence, $a + b + c = 12$ and $abc = 28$, find a, b, and c.
2) After inserting two numbers x and y between a and b, they form an arithmetic sequence. Find x and y.

Solution:

1) Since a, b, c form an arithmetic sequence, $2b = a + c$.

We have equation system $\begin{cases} b-a=c-b \\ a+b+c=12 \\ abc=28 \end{cases}$. Then $a = 1, b = 4, c = 7$.

2) To let the sequence $\{a, x, y, b\}$ be an arithmetic sequence, we have

$$\begin{cases} 2x=a+y \\ 2y=x+b \end{cases}$$

Then $x = \dfrac{2a+b}{3}$ and $y = \dfrac{a+2b}{3}$.

Thus the sequence $\{a, \dfrac{2a+b}{3}, \dfrac{a+2b}{3}, b\}$ is an arithmetic sequence.

Example 2.4.4   *Arithmetic Mean of Two Terms of A. S.*

Suppose the sequence $\{a_n\}$ is an arithmetic sequence $\{a_1, a_2, a_3, a_4, a_5, a_6, a_7, a_8, a_9\}$.

    1) If $a_1+a_2+a_3=15$ and $a_7+a_8+a_9=69$, find $a_4+a_5+a_6 = ?$.

    2) If $a_5 = 10$, find the sum $a_1+a_2+a_3+a_4+a_5+a_6+a_7+a_8+a_9 = ?$

Solution:

1) As $\{a_n\}$ is an arithmetic sequence, by (2.4.2) we have

and

$$
\begin{cases}
a_4 = \dfrac{a_1+a_7}{2}, \\[2mm]
a_5 = \dfrac{a_2+a_8}{2}, \\[2mm]
a_6 = \dfrac{a_3+a_9}{2}.
\end{cases}
$$

$$a_3+a_4+a_5 = \frac{a_1+a_2+a_3+a_7+a_8+a_9}{2} = \frac{15+69}{2} = 42$$

2) As $\{a_n\}$ is an arithmetic sequence, by (2.4.2) we have

$$a_5 = \frac{a_1+a_9}{2} = \frac{a_2+a_8}{2} = \frac{a_3+a_7}{2} = \frac{a_4+a_6}{2},$$

$$4 \cdot a_5 = \frac{a_1+a_2+a_3+a_4+a_6+a_7+a_8+a_9}{2}$$

Then $a_1+a_2+a_3+a_4+a_5+a_6+a_7+a_8+a_9 = 9\,a_5 = 90.$

▌2.5 Identifying Arithmetic Sequences

We can use the following laws to identify an arithmetic sequence.

The Laws of Identifying an Arithmetic Sequence

The sequence $\{a_n\}$ is an arithmetic sequence if any one of the following necessary and sufficient conditions is satisfied for all terms of the sequence.

1. Definition

$$a_n - a_{n-1} = \text{constant} \quad (\forall n, n \geqslant 2) \qquad (2.5.1)$$

2. Arithmetic mean

$$a_n = \frac{a_{n-1} + a_{n+1}}{2} \quad (\forall n, n \geqslant 2) \qquad (2.5.2)$$

3. The general term of $\{a_n\}$ is a linear function in terms of n.

$$a_n = pn + q \quad (p, q \text{ are constants}, \; p, q \neq 0)(2.5.3)$$

4. The sum S_n of the first n terms of the sequence $\{a_n\}$ is a quadratic function in terms of n without constant term.

$$S_n = pn^2 + qn \qquad (\forall n) \qquad (2.5.4)$$

▌**Proof**   *Law 2 - (2.5.2)*

Necessity:

Suppose arithmetic sequence $\{a_n\}$ has common difference d. Taking any three consecutive terms a_{n-1}, a_n, and a_{n+1}, we have following equation.

$$a_n - a_{n-1} = a_{n+1} - a_n = d$$

Thus

$$a_n = \frac{a_{n-1} + a_{n+1}}{2}.$$

Sufficiency:

If

$$a_n - \frac{a_{n-1} + a_{n+1}}{2},$$

we can write it as

$$a_n - a_{n-1} = a_{n+1} - a_n.$$

Let

$$a_n - a_{n-1} = a_{n+1} - a_n = d.$$

Because a_n is a random term, the equation above holds for all n, the sequence $\{a_n\}$ is an arithmetic sequence by the definition.

Conclusion

The sequence $\{a_n\}$ is an arithmetic sequence $\Longleftrightarrow^*$ $a_n = \dfrac{a_{n-1} + a_{n+1}}{2}$.

* "$\Longleftrightarrow$" *represents the necessary and sufficient relationship between two statements.*

■ Proof   Law 3 - (2.5.3)

Necessity

Because the sequence $\{a_n\}$ is an arithmetic sequence,

$$a_n = a_1 + (n-1)d = d \cdot n + (a_1 - d).$$

Let $p = d$ and $q = (a_1 - d)$,

$$a_n = pn + q.$$

The formula of the general term a_n is a linear function.

Sufficiency

Let the formula of the general term a_n be a linear function,

$$a_n = pn + q \qquad \text{(p and q are constants, $p, q \neq 0$)}$$

We get

$$a_{n-1} = p(n-1) + q$$

and

$$a_n - a_{n-1} = p.$$

By the definition, the sequence $\{a_n\}$ is an arithmetic sequence.

Conclusion

The sequence $\{a_n\}$ is an arithmetic sequence $\Longleftrightarrow$ $a_n = pn + q$.

■ Proof   Law 4 - (2.5.4)

Necessity

Because the sequence $\{a_n\}$ is an arithmetic sequence, we have

$$S_n = na_1 + \frac{n(n-1)d}{2} = \frac{d}{2}n^2 + \left(a_1 - \frac{d}{2}\right)n.$$

Let $p = d/2$ and $q = a_1 - d/2$,

$$S_n = pn^2 + qn.$$

It is a quadratic function without constant term in terms of n.

Sufficiency:

Suppose the formula of the sum S_n is a quadratic function without constant term in terms of n.

$$S_n = pn^2 + qn \qquad (p \text{ and } q \text{ are constants, } p, q \neq 0)$$

Then

$$S_{n-1} = p(n-1)^2 + q(n-1)$$
$$a_n = S_n - S_{n-1} = (2p)\cdot n + (q-p).$$

Therefore the formula of the general term a_n is a linear function in terms of n. By the law 3, the sequence $\{a_n\}$ is an arithmetic sequence.

Conclusion

The sequence $\{a_n\}$ is an arithmetic sequence $\iff S_n = pn^2 + qn$.

Example 2.5.1   *Identifying Arithmetic Sequences*

If $\{a_n\}$ is an arithmetic sequence with a_1 and common difference d, determine if the following sequences are arithmetic sequences.

1) $\{b_m\}$ consisting of all terms of $\{a_n\}$ remained after removing the first q terms of $\{a_n\}$.
2) $\{b_m\}$ consisting of all terms of $\{a_n\}$ with odd index n.
3) $b_n = p\,a_n + q$ (p and q are constants)

Solution:

1) $\{a_n\} = a_1, a_2, a_3, \overbrace{\cdots, a_q}^{q \text{ terms removed}}, \overbrace{a_{q+1}, a_{q+2}, \cdots, a_n,}^{\{b_m\}} \cdots.$

   Let $\{b_m\} = \{a_{q+1}, a_{q+2}, \cdots, a_{q+m-1}, a_{q+m}, \cdots\},$
   $$b_{m-1} = a_{q+m-1} = a_1 + (q+m-2)\cdot d$$
   $$b_m = a_{q+m} = a_1 + (q+m-1)\cdot d.$$

   Because $b_m - b_{m-1} = d$ is a constant, the sequence $\{b_m\}$ is an arithmetic sequence.

2) Let $\{b_m\} = \{a_1, a_3, a_5, \cdots, a_{2n-3}, a_{2n-1}, \cdots\},$
   $$b_{m-1} = a_{2n-3} = a_1 + (2n-4)\cdot d$$
   $$b_m = a_{2n-1} = a_1 + (2n-2)\cdot d.$$

   Because $b_m - b_{m-1} = 2d$ is a constant, the sequence $\{b_m\}$ is an arithmetic sequence.

3) From $b_n = p\,a_n + q$, $b_{n+1} = p\,a_{n+1} + q$.
   $$b_{n+1} - b_n = (pa_{n+1} + q) - (pa_n + q) = p(a_{n+1} - a_n)$$

   For the sequence $\{a_n\}$ is an arithmetic sequence, $a_{n+1} - a_n = d$ is a constant,
   $$b_{n+1} - b_n = p\cdot d$$

   is a constant. Therefore the sequence $\{b_n\}$ is an arithmetic sequence.

Example 2.5.2 *Identifying Arithmetic Sequences*

The sequence $\{a, b, c\}$ ($a>b>c>0$ and $d \neq 0$) is an arithmetic sequence, determine if the following sequences are arithmetic sequences.

　　1) $\{b + c, a + c, a + b\}$　　　　　　2) $\{bc, ca, ab\}$

Solution:

1) Since the sequence $\{a, b, c\}$ is an arithmetic sequence, we have

$$2b = a+c.$$

For the sequence $\{b+c, a+c, a+b\}$,

$$(b+c)+(a+b)=2b+a+c.$$

Because $2b=a+c$,　　$(b+c)+(a+b)=2(a+c)$

By Law 2, $(a + c)$ is the arithmetic mean of $(b + c)$ and $(a + b)$. Then the sequence $\{b+c, a+c, a+b\}$ is an arithmetic sequence.

2) Since the sequence $\{a, b, c\}$ is an arithmetic sequence, we have $2b=a+c$. Assume that the sequence $\{bc, ca, ab\}$ is an arithmetic sequences, then we have

$$2ac = bc+ab = b(a+c)$$

Substitute $b = \dfrac{a+c}{2}$ for b of above equation.

$$4ac = (a+c)^2$$
$$(a-c)^2 = 0$$

As the result $a=c$ is contrary to the condition $a>b>c>0$. Therefore the sequence $\{bc, ca, ab\}$ is not an arithmetic sequences.

Example 2.5.3 *Identifying Arithmetic Sequences*

1) If the sequence $\{bc, ac, ab\}$ ($a, b, c \neq 0$) is an arithmetic, determine if the sequence $\{\dfrac{b+c}{a}, \dfrac{c+a}{b}, \dfrac{a+b}{c}\}$ is an arithmetic sequences.

2) If the sequence $\{a^2, b^2, c^2\}$ is an arithmetic sequence, prove the sequence $\{\dfrac{1}{b+c}, \dfrac{1}{a+c}, \dfrac{1}{a+b}\}$ is an arithmetic sequence.

Solution:

1) We check if these three terms meet Law 2.

　　For $2ac=bc+ab=b(a+c)$, we have

$$\frac{b+c}{a}+\frac{a+b}{c}=\frac{(a+c)^2}{ac}=2\frac{a+c}{b}.$$

By Law 2, the sequence $\{\dfrac{b+c}{a},\dfrac{c+a}{b},\dfrac{a+b}{c}\}$ is an arithmetic sequence.

2) We have $\dfrac{1}{b+c}+\dfrac{1}{a+b}=\dfrac{a+2b+c}{(a+b)(b+c)}=\dfrac{a+2b+c}{ab+ac+b^2+bc}.$

Because the sequence $\{a^2, b^2, c^2\}$ is an arithmetic sequence we have $a^2+c^2=2b^2.$ Then the formula above becomes

$$\dfrac{a+2b+c}{ab+ac+\dfrac{a^2+c^2}{2}+bc}=\dfrac{2(a+2b+c)}{(a+2b+c)(a+c)}=\dfrac{1}{b+c}+\dfrac{1}{a+b}=\dfrac{2}{a+c}.$$

By Law 2, the sequence $\{\dfrac{1}{b+c},\dfrac{1}{a+c},\dfrac{1}{a+b}\}$ is an arithmetic sequence.

Example 2.5.4   *Identifying Arithmetic Sequences*

If the sequence $\{a_n\}$ is defined by $a_n=3\,n-1$, determine if $\{a_n\}$ is an arithmetic sequence. (Use four methods)

Solution:

- Law 1 (the definition)

  Since $a_n=3\,n-1$ and $a_{n-1}=3(n-1)-1=3\,n-4,$

  $$d=a_n-a_{n-1}=3.$$

  Because $d = 3$ is constant, by Law 1, $\{a_n\}$ is an arithmetic sequence.

- Law 2 (arithmetic mean)

  Since $a_n=3\,n-1, a_{n-1}=3(n-1)-1=3\,n-4$ and $a_{n+1}=3(n+1)-1=3\,n+2,$

  $$\dfrac{a_{n-1}+a_{n+1}}{2}=3\,n-1=a_n.$$

  Then $\{a_n\}$ is an arithmetic sequence by Law 2.

- Law 3 (general term a_n):

  $a_n=3\,n-1$ is a linear function, thus the sequence $\{a_n\}$ is an arithmetic sequence.

- Law 4 (sum S_n)

  Since $a_1=3\cdot1-1=2,$

  $$S_n=\dfrac{n(a_1+a_n)}{2}=\dfrac{n(2+3n-1)}{2}=\dfrac{3}{2}n^2+\dfrac{1}{2}n.$$

  The formula of S_n is a quadratic function without constant term. By Law 4, $\{a_n\}$ is an arithmetic sequence.

Example 2.5.5   *Identifying Arithmetic Sequences*

If $S_n=2n^2-n+2$, determine if the sequence $\{a_n\}$ is an arithmetic sequence. (Use four methods)

Solution:

We can get the general term a_n from the sum S_n.

$$a_n = \begin{cases} S_1=2\cdot1^2-1+2=3 & (n = 1) \\ S_n-S_{n-1}=2n^2-n+2-[2(n-1)^2-(n-1)+2]=2n-3 & (n \geqslant 2) \end{cases}$$

- Law 1 (the definition)

  Since $a_1=3$, $a_2=1$ and $a_3=3$, we have
  $$a_2-a_1=1-3=-2$$
  and
  $$a_3-a_2=3-1=2.$$
  Because $a_2-a_1 \neq a_3-a_2$, the sequence $\{a_n\}$ is not an arithmetic sequence even though $a_n-a_{n-1}=2$ $(n \geqslant 3)$ is a constant for the rest of the sequence.

- Law 2 (arithmetic mean)

  Since $a_1+a_3 \neq 2a_2$, the sequence $\{a_n\}$ is not an arithmetic sequence.

- Law 3 (general term a_n)

  The general term a_n is a piecewise function and when using the general term we have $a_1=2\cdot n-3 \neq 3$. The sequence $\{a_n\}$ is not an arithmetic sequence.

- Law 4 (the sum S_n)

  Because the formula $S_n=2n^2-n+2$ has a constant term, the sequence $\{a_n\}$ is not an arithmetic sequence.

Example 2.5.6   *Identifying Arithmetic Sequences*

If $S_n=6n^2-3n$, determine if the sequence $\{a_n\}$ is an arithmetic sequence. (Use four methods)

Solution:

We can get the general term a_n from the sum S_n.

$$a_1=S_1=6\cdot1^2-3=3 \qquad\qquad (n = 1)$$
$$a_n=S_n-S_{n-1}=6n^2-3n-[6(n-1)^2-3(n-1)]=12n-9 \qquad (n \geqslant 2)$$

- Law 1 (the definition)

  Since $a_n=12n-9$, $a_{n+1}=12(n+1)-9$, we get
  $$a_{n+1}-a_n=12.$$
  Because $a_{n+1}-a_n=12$ is a constant, $\{a_n\}$ is an arithmetic sequence.

- Law 2 (arithmetic mean)

Since $a_{n-1}=12(n-1)-9$, $a_{n+1}=12(n+1)-9$,

$$2a_n=a_{n+1}+a_n,$$

the sequence $\{a_n\}$ is an arithmetic sequence.

- Law 3 (general term a_n)
  Because the general term $a_n=12n-9$ is a linear function, the sequence $\{a_n\}$ is an arithmetic sequence by Law 3.

- Law 4 (sum S_n)
  Because the sum of the first n terms of the sequence $\{a_n\}$, $S_n=6n^2-3n$, is a quadratic function without constant term in terms of n, $\{a_n\}$ is an arithmetic sequence by Law 4.

Example 2.5.7   *Identifying Arithmetic Sequences*

Suppose $a_1=1$ and $a_n=-3S_n \cdot S_{n-1}$ $(n\geqslant2)$.

1) Determine if the sequence $\{\dfrac{1}{S_n}\}$ is an arithmetic sequence.

2) Find the general term a_n.

Solution:

1) Because $a_n=S_n-S_{n-1}$, $S_n-S_{n-1}=-3S_n \cdot S_{n-1}$.

$$\frac{1}{S_n}-\frac{1}{S_{n-1}}=3$$

By Law 1 (2.5.1), the sequence $\{\dfrac{1}{S_n}\}$ is an arithmetic sequence.

2) We use the formula $a_n=S_n-S_{n-1}$ to find the general term a_n because we can not assume that the sequence $\{a_n\}$ is an arithmetic sequence. For the sequence $\{\dfrac{1}{S_n}\}$, $\dfrac{1}{S_1}=\dfrac{1}{a_1}=1$ and the common difference $d = 3$, the general term becomes

$$\frac{1}{S_n}=\frac{1}{S_1}+(n-1)\cdot3=3n-2.$$

We have $S_n=\dfrac{1}{3n-2}$ and $S_{n-1}=\dfrac{1}{3(n-1)-2}=\dfrac{1}{3n-5}$.

$$a_n=S_n-S_{n-1}=\frac{-3}{(3n-2)(3n-5)}.$$

$$a_n = \begin{cases} 1 & (n=1) \\ \dfrac{-3}{(3n-2)(3n-5)} & (n\geqslant2). \end{cases}$$

Example 2.5.8   *Identifying Arithmetic Sequences*

Suppose that the sequences $\{b_n\}$ is an arithmetic sequence with common difference $d \neq 0$ and

$$a_1 + 2a_2 + 3a_3 + \cdots + na_n = (1 + 2 + 3 + \cdots + n)b_n.$$

If $a_1 \neq 0$, prove the sequence $\{a_n\}$ is an arithmetic sequence.

Solution:

$$\frac{n(n+1)}{2}b_n = a_1 + 2a_2 + 3a_3 + \cdots + na_n$$

$$\frac{n(n-1)}{2}b_{n-1} = a_1 + 2a_2 + 3a_3 + \cdots + (n-1)a_{n-1} \quad (n \geqslant 2).$$

$$\frac{n(n+1)}{2}b_n - \frac{n(n-1)}{2}b_{n-1} = na_n$$

$$a_n = \frac{n+1}{2}b_n - \frac{n-1}{2}b_{n-1}$$

As $b_{n-1} = b_n - d$ and $b_n = b_1 + (n-1)d$,

$$a_n = b_n + \frac{n-1}{2}d = b_1 + (n-1)\frac{3}{2} \cdot d.$$

By Law 3 the sequence $\{a_n\}$ is an arithmetic sequence whose $a_1 = b_1$, and the common difference is $\frac{3}{2}d$.

■ 2.6 Properties of Arithmetic Sequences

▶ Property 1

Suppose that $\{a_n\}$ is an arithmetic sequence with the common difference d.

If $\begin{cases} d = 0 \\ d > 0 \\ d < 0 \end{cases}$, $\{a_n\}$ is $\begin{cases} \text{a constant sequence.} \\ \text{an increasing sequence and } a_1 \text{ is the minimum.} \\ \text{a decreasing sequence and } a_1 \text{ is the maximum.} \end{cases}$ (2.6.1)

Example 2.6.1   *Types of Arithmetic Sequences*

Discuss the monotony of the following sequence $\{a_n\}$.

   1) $\lambda, \lambda, \lambda, \cdots$ 2) $4, 10, 16, 22, \cdots$ 3) $12, 10, 8, 6, \cdots$

Solution:

1) Because $d = a_n - a_{n-1} = \lambda - \lambda = 0$, $\{a_n\}$ is a constant sequence.
2) $a_n = 6n - 2$, $d = a_n - a_{n-1} = 6 > 0$, $\{a_n\}$ is an increasing sequence.
3) Since $a_n = 14 - 2n$, $d = a_n - a_{n-1} = -2 < 0$, $\{a_n\}$ is a decreasing sequence.

▶ Property 2

The sequence $\{b_n\}$ is an arithmetic sequence with common difference $p \cdot d_a$ if
   1) the sequence $\{a_n\}$ is an arithmetic sequence whose common difference is d_a.
   2) b_n meets one of equations below (p and q are constants, $p, q \neq 0$).

$$b_n = p \cdot a_n \qquad\qquad (2.6.2)$$
$$b_n = p \cdot a_n + q \qquad\qquad (2.6.3)$$

■ Proof   *Property 2 - (2.6.3)*

Suppose that the sequence $\{a_n\}$ is an arithmetic sequence with common difference d_a, p and q are constants.

Let $b_n = p\, a_n + q$ and $b_{n-1} = p\, a_{n-1} + q$, then $b_n - b_{n-1} = p \cdot (a_n - a_{n-1}) = p \cdot d_a$.

As $p \cdot d_a$ is a constant, the sequence $\{b_n\}$ is an arithmetic sequence.

Example 2.6.2   *Application of Property 2*

Determine if two sequences $a_n = 2n+5$ and $b_n = a_n + 3$ are arithmetic sequences.

Solution:

By the law 3 (2.5.3), the sequence $\{a_n\}$ is an arithmetic sequence.
By the property 2 (2.6.3), the sequence $\{b_n\}$ is an arithmetic sequence.

▶ Property 3

If the sequences $\{a_n\}$ and $\{b_n\}$ are arithmetic sequences with common difference d_a and d_b respectively, then the sequence

$$\{pa_n + qb_n\} \quad (p \text{ and } q \text{ are constants}, p, q \neq 0) \qquad (2.6.4)$$

is an arithmetic sequence with common difference $p d_a + q d_b$.

Proof   *Property 3 - (2.6.4)*

Suppose that c_n is the general term of the sequence $\{c_n\}$ and $c_n = p a_n + q b_n$.

We have

$$\begin{cases} c_n = p a_n + q b_n = p a_1 + q b_1 + (n-1)(p d_a + q d_b) \\ c_{n-1} = p a_{n-1} + q b_{n-1} = p a_1 + q b_1 + (n-2)(p d_a + q d_b) \end{cases}$$

Since $c_n - c_{n-1} = p d_a + q d_b$ is a constant, the sequence $\{p a_n + q b_n\}$ is an arithmetic sequence by the definition of arithmetic sequences.

Example 2.6.3   *Application of Property 3*

If $a_n = 2n + 3$ and $b_n = 4n$ and $c_n = 2a_n + 3b_n$ find c_{10}.

Solution:

Since $c_n = 2 a_n + 3 b_n = 2(2 n+3) + 3(4 n) = 16 n + 6$, $c_{10} = 16 \cdot 10 + 6 = 166$.

Example 2.6.4   *Application of Property 3*

Suppose that the sequences $\{a_n\}$, $\{b_n\}$, and $\{c_n\}$ are arithmetic sequences. If $a_n = 2n+3$, $c_n = 3 a_n + 6 b_n$, $c_1 = 27$, and $c_3 = 75$, find b_n.

Solution:

For the sequence $\{c_n\}$ is an arithmetic sequence, $c_n = c_1 + (n-1) d_c$. We obtain $c_3 = 27 + (3-1) d_c = 75$ and $d_c = 24$.

Then $c_n = 27 + (n-1) \cdot 24 = 24 n + 3.$

Let

$$c_n=3a_n+6b_n=24n+3 \tag{a}$$

As the sequence $\{b_n\}$ is an arithmetic sequence, its general term must be a linear function in terms of n. Let $b_n=kn+\lambda$. Substitute $24n+3$ and $kn+\lambda$ for a_n and b_n of (a) respectively.

$$3(2n+3)+6(kn+\lambda)=24n+3$$
$$6(1+k)n+(9+6\lambda)=24n+3$$

After comparing two sides of above equation we have

$$\begin{cases} 6(1+k)=24 \\ 9+6\lambda=3 \end{cases}.$$

We obtain $k=3$ and $\lambda=-1$ then $b_n=3n-1$.

▶ Property 4

If
- the sequence $\{a_n\}$ is an arithmetic sequence and
- the sequence $\{b_m\}$ $(b_m \in N^+, b_m \geqslant 1)$ is an arithmetic sequence of positive integers

then the sequence

$$\{a_{b_1}, a_{b_2}, \cdots, a_{b_m}, \cdots\} \tag{2.6.5}$$

is an arithmetic sequence with common difference $d_a \cdot d_b$
(d_a and d_b are common difference for $\{a_n\}$ and $\{b_m\}$ respectively.)

Proof   *Property 4 - (2.6.5)*

Notice the difference of the subscriptions of any two consecutive terms of the sequence $\{a_{b_1}, a_{b_2}, \cdots, a_{b_m}, \cdots\}$ is d_b.

$$a_{b_m}=a_{b_1}+(b_m-b_1)\cdot d_a$$
$$a_{b_{m-1}}=a_{b_1}+(b_{m-1}-b_1)\cdot d_a$$
$$=a_{b_1}+(b_m-d_b-b_1)\cdot d_a$$

Then the common difference of the sequence $\{a_{b_1}, a_{b_2}, \cdots, a_{b_m}, \cdots\}$ becomes

$$a_{b_m}-a_{b_{m-1}}=d_a \cdot d_b.$$

The sequence is an arithmetic sequence.

Example 2.6.5   *Application of Property 4*

Suppose $a_n = 3n + 8$, $b_n = 2n + 5$ ($n \in N^+$, $n \geqslant 1$), and $\{c_n\}$: $\{a_{b_1}, a_{b_2}, \cdots, a_{b_m}, \cdots\}$.

1) Prove the sequence $\{c_n\}$ is an arithmetic sequence.
2) Find c_{100} and the subscription of corresponding term a_n of the sequence $\{a_n\}$.
3) Let B_n and C_n be the sum of the first n terms of the sequence $\{b_n\}$ and $\{c_n\}$ respectively. If $C_k = 560$, find B_k.

Solution:

1) Because both a_n and b_n are linear functions in terms of n, the sequence $\{a_n\}$ and $\{b_n\}$ are arithmetic sequences by the law 3. Since n is a positive integer, $b_n = 2n + 5$ must be a positive integer too. By Property 4, the sequence $\{c_n\}$ is an arithmetic sequence.

2) For the sequence $\{c_n\}$ is an arithmetic sequence, we have $c_n = c_1 + (n-1)d_c$ and $d_c = d_a \cdot d_b = 6$.
   Since $b_1 = 7$, $c_1 = a_{b_1} = a_7 = 3 \cdot 7 + 8 = 29$,

$$c_n = 29 + 6(n-1) = 6n + 23.$$

We obtain

$$c_{100} = 6 \cdot 100 + 23 = 623.$$

3) Let

$$C_k = k \cdot c_1 + \frac{k(k-1)d_c}{2}$$

$$= k \cdot 29 + \frac{k(k-1) \cdot 6}{2} = 560,$$

then $k = 10$. We obtain

$$B_k = k\, b_1 + \frac{k(k-1)d_b}{2} = 160.$$

▶ Property 5

Suppose a sequence $\{a_n\}$ is an arithmetic sequence with common difference $d \neq 0$. If
- The sequence $\{h_m\}$: $\{h_1, h_2, \cdots, h_m\}$ and $\{k_m\}$: $\{k_1, k_2, \cdots, k_m\}$ are two finite sequences of positive integers and both have m terms.
- $h_1 + h_2 + \cdots + h_m = k_1 + k_2 + \cdots + k_m$.

then

$$a_{h_1} + a_{h_2} + \cdots + a_{h_m} = a_{k_1} + a_{k_2} + \cdots + a_{k_m} \qquad (2.6.6)$$

| ■ **Proof** | *Property 5 - (2.6.6)* |

Suppose

    a) $\{h_m\}:\{h_1, h_2, \cdots, h_m\}$ and $\{k_m\}:\{k_1, k_2, \cdots, k_m\}$ are two sequences of positive integers and both have m terms.

    b) $h_1 + h_2 + \cdots + h_m = k_1 + k_2 + \cdots + k_m$.

Because the sequence $\{a_n\}$ is an arithmetic sequence we have

$$\begin{cases} a_{h_1} = a_1 + (h_1 - 1) \cdot d \\ \cdots \\ a_{h_m} = a_1 + (h_m - 1) \cdot d \end{cases}$$

Add all m equations above, we have

$$a_{h_1} + a_{h_2} + \cdots + a_{h_m} = ma_1 + (h_1 + h_2 + \cdots + h_m - m)d$$

In the same way, $\quad a_{k_1} + a_{k_2} + \cdots + a_{k_m} = ma_1 + (k_1 + k_2 + \cdots + k_m - m)d$.

Therefore $\quad a_{h_1} + a_{h_2} + \cdots + a_{h_m} = a_{k_1} + a_{k_2} + \cdots + a_{k_m}$.

| **Example 2.6.6** | *Application of Property 5* |

1) If the sequence $\{a_n\}$ is an arithmetic sequence, prove $S_n = \dfrac{n(a_1 + a_n)}{2}$.

2) If the sequence $\{a_n\}$ is an arithmetic sequence and
$a_2 + a_k + a_{15} = a_4 + a_7 + a_{12} = 20$, find subscription k and $a_k + a_8 + a_9 = ?$

Solution:

1) The sequences $\{1, n\}$ and $\{2, n-1\}$ have two terms and $1 + n = 2 + (n-1)$. By Property 5, we have $a_1 + a_n = a_2 + a_{n-1}$. In the same way we can prove

$$a_2 + a_{n-1} = a_3 + a_{n-2}, a_2 + a_{n-1} = a_3 + a_{n-2}, \cdots$$

and we have $n/2$ equations:

$$\left. \begin{array}{l} a_2 + a_{n-1} = a_1 + a_n \\ a_3 + a_{n-2} = a_2 + a_{n-1} \\ \cdots \end{array} \right\} \quad (n/2 \text{ equations})$$

Add all equations, then

$$S_n = a_1 + a_2 + a_3 + \cdots + a_n = \frac{n}{2}(a_1 + a_n).$$

2) Let $k + 2 + 15 = 4 + 7 + 12$, $k = 6$. Since $a_6 + a_8 + a_9 = a_4 + a_7 + a_{12}$,

$$a_k + a_8 + a_9 = 20.$$

Example 2.6.7   Application of Property 5

If the sequence $\{a_n\}$ is an arithmetic sequence, $a_1+a_2+a_3=18$, and $a_{15}+a_{16}+a_{17}=102$, find a_9 and S_9.

Solution:

Let
$$(a_1+a_2+a_3)+(a_{15}+a_{16}+a_{17})=18+102=120.$$

For the sum of the subscripts $1+2+3+15+16+17=54=6\cdot9$, by Property 5,

let
$$6\cdot a_9=a_1+a_2+a_3+a_{15}+a_{16}+a_{17}=120.$$

Therefore
$$a_9=a_1+8d=20 \tag{a}$$

On the other hand, $a_1+a_2+a_3=18$ can be written as

$$a_1+(a_1+d)+(a_1+2d)=3a_1+3d=18. \tag{b}$$

Solve the equation system of (a) and (b)
$$\begin{cases} a_1+8d=20 \\ a_1+d=6 \end{cases}$$

Then $a_1=4$ and $d=2$.

Thus
$$S_9=\frac{9(a_1+a_9)}{2}=\frac{9(4+20)}{2}=108.$$

Example 2.6.8   Application of Property 5

If the sequence $\{a_n\}$ is an arithmetic sequence and $a_3-a_5+4\cdot a_6-a_7+a_9=8$. find
1) $a_2+a_4+a_{12}=$?   2) $S_{11}=$?

Solution:

By Property 5, $(a_3+a_9)+4a_6-(a_5+a_7)=8$, $2a_6+4a_6-2a_6=8$ then $a_6=2$.

1) $a_2+a_4+a_{12}=3\cdot a_6=6$.

2) Since $a_1+a_{11}=2a_6=4$, $S_{11}=\dfrac{11(a_1+a_{11})}{2}=\dfrac{11\cdot4}{2}=22$.

Example 2.6.9   Application of Property 5

If the sequence $\{a_n\}$ is an arithmetic sequence, $a_4+a_5+a_6+a_7=74$, and $a_2+a_3+a_6+a_8=65$, find the common difference d.

Solution:

As
$$\begin{cases} a_4+a_5+a_6+a_7=(a_4+a_7)+(a_5+a_6)=2a_4+2a_7=74 \\ a_2+a_3+a_6+a_8=(a_2+a_6)+(a_3+a_8)=2a_4+(a_4+a_7)=3a_4+a_7=65 \end{cases}$$

we have
$$\begin{cases} a_4+a_7=37 \\ 3a_4+a_7=65 \end{cases}$$

After solving the above we get $a_4 = 14$ and $a_7 = 23$. By the definition of arithmetic sequences, we have

$$\begin{cases} a_4 = a_1 + 3 \cdot d = 14 \\ a_7 = a_1 + 6 \cdot d = 23 \end{cases}$$

Therefore the common difference $d = 3$.

▶ Property 6

The Extreme for the Sum of an Arithmetic Sequence

Suppose that S_n is the sum of the first n terms of an arithmetic sequence $\{a_n\}$ whose first term is a_1 and common difference is d.

1) a_1 and d have the same sign,
   - If $a_1 \geqslant 0$ and $d > 0$, $\{a_n\}$ is an increasing sequence. S_1 is the minimum in $\{S_n\}$.
   - If $a_1 \leqslant 0$ and $d < 0$, $\{a_n\}$ is a decreasing sequence. S_1 is the maximum in $\{S_n\}$.

2) a_1 and d have different sign,
   - If $a_1 > 0$ and $d < 0$, $\{a_n\}$ is a decreasing sequence, there is the maximum in $\{S_n\}$ where $a_n \geqslant 0$ and $a_{n+1} < 0$.
   - If $a_1 < 0$ and $d > 0$, $\{a_n\}$ is an increasing sequence, there is the minimum in $\{S_n\}$ where $a_n \leqslant 0$ and $a_{n+1} > 0$.

As an arithmetic sequence is a linear discrete function we can discuss it by its monotony.

1) When a_1 and d have the same sign, $\{a_n\}$ does not cross the n- axis.
   - **$a_1 \geqslant 0$ and $d > 0$**

     The sequence $\{a_n\}$ is an increasing sequence. The first term a_1 is above the n-axis. Since $a_n \geqslant 0$, the sum S_n is accumulated by positive terms only. Thus $S_1 = a_1$ is the minimum.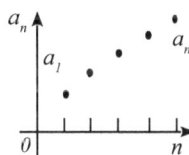

   - **$a_1 \geqslant 0$ and $d < 0$**

     The sequence $\{a_n\}$ is a decreasing sequence. The first term a_1 is below the n-axis. Since $a_n \leqslant 0$, the sum S_n is accumulated by negative terms only. Then $S_1 = a_1$ is the maximum.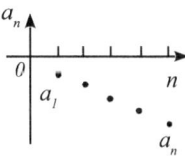

2) When a_1 and d have different sign, $\{a_n\}$ will crosses the n-axis.

- **$a_1 > 0$ and $d < 0$**
  The sequence $\{a_n\}$ is a decreasing sequence and the first term a_1 is above the n-axis. When $a_n \geqslant 0$ and $a_{n+1} < 0$, the sequence $\{a_n\}$ will cross the n-axis to enter the negative zone. S_n is accumulated by positive terms only above the n-axis then negative terms below the n-axis. Then S_n will reach its maximum when $a_n \geqslant 0$ and $a_{n+1} < 0$.

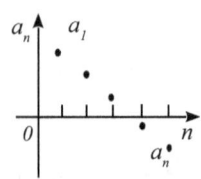

- **$a_1 < 0$ and $d > 0$**
  The sequence $\{a_n\}$ is an increasing sequence and the first term a_1 is below the n-axis. When $a_n \leqslant 0$ and $a_{n+1} > 0$. the sequence $\{a_n\}$ will cross the n-axis enter the positive zone. S_n is accumulated by negative terms only below the n-axis then positive terms above the n-axis. Then S_n will reach its minimum when $a_n \leqslant 0$ and $a_{n+1} > 0$.

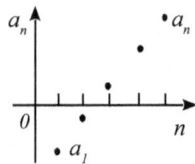

From Property 1 of arithmetic sequences, an arithmetic sequence is a monotone sequence if its common difference $d \neq 0$. The following methods are used to find the extreme of the sum S_n of the arithmetic sequence $\{a_n\}$ by the monotone of arithmetic sequences.

1) To find the extreme of the sum S_n use the following two methods
   - Find the extreme of the quadratic function of S_n in terms of n.
     The extreme of the sum S_n can be obtained by finding the extreme of the quadratic function of S_n in terms of n.

$$S_n = n a_1 + \frac{n(n-1)}{2} d = \frac{d}{2} n^2 + \left(a_1 - \frac{d}{2}\right) n$$

   - Find where the term a_n changes its sign
     Use Property 6 arithmetic sequences to find the extreme of an arithmetic sequence because the extreme of the sum S_n occurs where $\{a_n\}$ crosses the n-axis and the term a_n changes its sign.

2) To find the maximum of n when $S_n > 0$ or $S_n < 0$
   - Let $S_n > 0$ to find the maximum of n where $S_n > 0$.
   - Let $S_n < 0$ to find the maximum of n where $S_n < 0$.

Example 2.6.10 *The Extreme of the Sum of Arithmetic Sequences*

The sequence $\{a_n\}$ is an arithmetic sequence and $a_1 > 0$, if

1) $S_5 = S_{17}$, find n when S_n reach its maximum.
2) $2\,a_{10} = 3\,a_{17}$, find n when S_n reach its maximum.

Solution:

1) Since $S_5 = 5\,a_1 + 10 \cdot d$ and $S_{17} = 17\,a_1 + 136 \cdot d$,

let
$$5\,a_1 + 10 \cdot d = 17\,a_1 + 136 \cdot d.$$

We obtain
$$d = -\frac{2}{21}\,a_1.$$

Then
$$S_n = n\,a_1 + \frac{n(n-1)}{2}d$$
$$= n\,a_1 + \frac{n(n-1)}{2}\left(-\frac{2}{21}a_1\right)$$
$$= \frac{121}{21}\,a_1 + \left(-\frac{a_1}{21}\right)(n-11)^2.$$

Because $a_1 > 0$ and $d < 0$, the sequence $\{a_n\}$ is a decreasing sequence. To let S_n reach the maximum we eliminate the second part of the formula above by setting $n = 11$, which is always negative. So $S_n = \frac{121}{21}\,a_1$ is the maximum when $n = 11$.

2) Let $2\,(a_1 + 9\,d) = 3\,(a_1 + 16d)$. We have $d = -\frac{a_1}{30}$ and

$$a_n = a_1 + (n-1)\left(-\frac{a_1}{30}\right) = (31 - n)\frac{a_1}{30}.$$

Because $a_1 > 0$ and $d < 0$, the sequence $\{a_n\}$ is a decreasing sequence and S_n will reach its maximum when the sequence $\{a_n\}$ crosses the n-axis, where $a_n \geq 0$ and $a_{n+1} < 0$, to enter negative zone.

Let
$$a_n = (31 - n)\frac{a_1}{30} = 0.$$

Since $\frac{a_1}{30} > 0$, $31 - n = 0$. Therefore S_n reaches its maximum when $n = 31$.

Example 2.6.11   *The Extreme of the Sum of Arithmetic Sequences*

1) Suppose that $\{a_n\}$ is an arithmetic sequence, $a_1=15$, $a_8>0$, $a_9<0$, and the common difference d is an integer. Find the maximum S_n and the maximum n when $S_n \geqslant 0$.

2) The sequence $\{a_n\}$ is an arithmetic sequence and $a_n=3\,n-16$. Find n when S_n reaches the minimum S_{min}.

3) Suppose that the sequence $\{a_n\}$ is an arithmetic sequence, $a_1+a_4+a_7=9$, and $a_2+a_5+a_8=21$, find n when S_n reaches its minimum S_{min}.

Solution:

1) Since
$$\begin{cases} a_1+7d>0 \\ a_1+8d<0 \end{cases},$$

we obtain
$$-2\frac{1}{7}<d<-1\frac{7}{8}$$

and take $d=-2$.

As $a_1>0$ and $d<0$, the sequence $\{a_n\}$ is a decreasing sequence. Because $a_8>0$ and $a_9<0$, $\{a_n\}$ will crosses the n-axis between $n=8$ and $n=9$. Then S_n will reach its maximum at $n=8$.

$$S_{max}=S_8=8\cdot15+\frac{8(8-1)\cdot(-2)}{2}=64.$$

Let
$$S_n=15\cdot n+\frac{n(n-1)\cdot(-2)}{2}=n(16-n)\geqslant 0.$$

Since $n>0$, let $16-n\geqslant 0$, i.e. $n\leqslant 16$ and take $n_{max}=16$.

2) Since the sequence $\{a_n\}$ is an arithmetic sequence,
$$a_{n+1}-a_n=d.$$

For $a_n=3\,n-16$ and $a_{n+1}=3(n+1)-16$, $d=a_{n+1}-a_n=3$. Because $a_1=-13<0$ and $d>0$, the sequence $\{a_n\}$ is an increasing sequence. It will cross the n-axis from the bottom of the n-axis where S_n will reach its minimum.

Let
$$\begin{cases} a_n=3\,n-16<0 \\ a_{n+1}=3(n+1)-16>0 \end{cases}.$$

We obtain $4\frac{1}{3}<n<5\frac{1}{3}$ and take $n=5$. When $n=5$, S_n will reach its minimum

$$S_{min}=S_5=5\cdot(-13)+\frac{5(5-1)\cdot3}{2}=-35.$$

3) By the theorem (1.2.9) we have $\begin{cases} a_1+a_7=2\,a_4 \\ a_2+a_8=2\,a_5 \end{cases}$

Then $a_4=3$ and $a_5=7$. Because the sequence $\{a_n\}$ is an arithmetic sequence

$$\begin{cases} a_4=a_1+3\cdot d=3 \\ a_5=a_1+4\cdot d=7 \end{cases}.$$

We obtain $d=4$ and $a_1=-9$. The general term becomes

$$a_n=-9+(n-1)\cdot 4=4n-13.$$

The sequence $\{a_n\}$ is an increasing sequence. It will cross the n-axis from the bottom of the n-axis where S_n will reach its minimum. Find n at which a_n will switch its sign.

Let

$$\begin{cases} a_n=4n-13<0 \\ a_{n+1}=4(n+1)-13>0 \end{cases}$$

We have $\dfrac{9}{4}<n<\dfrac{13}{4}$ and take $n=3$.

Therefore

$$S_{min}=S_3=3\cdot(-9)+\frac{3(3-1)\cdot 4}{2}=-15.$$

▶ Property 7

Sequence of the Sums of Term Groups of Arithmetic Sequences

If

1) The sequence $\{a_n\}$ is an arithmetic sequence with common difference d.

2) The term groups of the sequence $\{a_n\}$ are constructed by the rules in the section 1.1.7 and denoted by the following form.

$$\begin{cases} \cdots \\ \{g_{k-1}\}:\{a_{n-p-m},\cdots,a_{n-p-1}\}, \\ \{g_k\}: \{a_n,\cdots,a_{n+m-1}\} \qquad (k,m,p\in N^*, 1<m<n, p\geqslant 0) \\ \{g_{k+1}\}:\{a_{n+p+m},\cdots,a_{n+p+2m-1}\} \\ \cdots \end{cases}$$

$\{G_k\}$ denotes the sequence $\{\ \{g_1\}, \{g_2\}, \cdots, \{g_k\}, \cdots\ \}$.

3) T_k is the sum of all terms of the kth term group $\{g_k\}$ of the sequence $\{G_k\}$, that is $T_k=a_n+a_{n+1}+\cdots+a_{n+m-2}+a_{n+m-1}$.

4) $\{T_k\}$ denotes the sequence $\{T_1, \cdots, T_k, \cdots\}$.

then $\{T_k\}$ is an arithmetic sequence with common difference $m(m+p)d$.

▌ Proof   *Property 7 of Arithmetic Squence*

Suppose that $\{a_n\}$ is an arithmetic sequence with common difference d. Term groups are built within the sequence $\{a_n\}$ by the rules in section 1.1.7. Let the sequence $\{G_k\}$ contain these term groups.

$$\{G_k\}:\{\ \{g_1\}\ ,\ \{g_2\}\ ,\ \cdots,\ \{g_k\},\ \cdots\ \}$$

be the sequence containing all term groups of $\{a_n\}$.

Then $\{g_k\}:\ \{a_n,\cdots,a_{n+m-1}\}$

and $\{g_{k+1}\}:\{a_{n+p+m},\cdots,a_{n+p+2m-1}\}.$

Let T_k be the sum of all terms of the kth term group $\{g_k\}$, we obtain the sequence $\{T_k\}$.

$$\{T_k\}:\{T_1,\ \cdots,\ T_k,\ T_{k+1},\ \cdots\}$$

Where $T_k\ =a_n+\cdots+a_{n+m-1}$

$$T_{k+1}=a_{n+p+m}+\cdots+a_{n+p+2m-1}.$$

Because $\{a_n\}$ is an arithmetic sequence we have

$$T_k=a_n+\cdots+a_{n+m-1}=S_{n+m-1}-S_{n-1}$$

$$=[(n+m-1)\cdot a_1+\frac{(n+m-1)(n+m-2)\cdot d}{2}]-[(n-1)\cdot a_1+\frac{(n-1)(n-2)\cdot d}{2}]$$

$$=m\,a_1+\frac{m(2n+m-3)d}{2}$$

$$T_{k+1}=a_{n+p+m}+\cdots+a_{n+p+2m-1}=S_{n+p+2m-1}-S_{n+p+m-1}$$

$$=[(n+p+2m-1)\cdot a_1+\frac{(n+p+2m-1)(n+p+2m-2)\cdot d}{2}].$$

$$-[(n+p+m-1)\cdot a_1+\frac{(n+p+m-1)(n+p+m-2)\cdot d}{2}]$$

$$=m\,a_1+\frac{m(2n+2p+3m-3)d}{2}.$$

Since $T_{k+1}-T_k=m(m+p)\,d$

is a constant the sequence $\{T_k\}$ is an arithmetic sequence by the definition of arithmetic sequences.

Example 2.6.12   *Sequence of the Sums of Term Groups of an A.S.*

If the sequence $\{a_n\}$ is an arithmetic sequence, $a_3+a_4+a_5=30$ and $a_8+a_9+a_{10}=60$, find

1) $a_{13}+a_{14}+a_{15}=?$ 2) $a_{23}+a_{24}+a_{25}=?$ 3) $a_{44}=?$

Solution:

1) Construct the term groups with $m = 3$ and $p = 2$ and we have

and
$$\begin{cases} \{g_1\}:\{a_{3,}\,a_{4,}\,a_5\}, \\ \{g_2\}:\{a_{8,}\,a_{9,}\,a_{10}\}, \\ \{g_3\}:\{a_{13,}\,a_{14,}\,a_{15}\}. \end{cases}$$

Let
$$\begin{cases} T_1=a_3+a_4+a_5=30 \\ T_2=a_8+a_9+a_{10}=60 \\ T_3=a_{13}\mid a_{14}\mid a_{15} \end{cases}$$

and
$$\{T_k\}:\{T_1, T_2, T_3, \cdots\}.$$

Then the sequence $\{T_k\}$ is an arithmetic sequence by Property 7. We have

$$2T_2=T_1+T_3.$$

Thus
$$T_3=a_{13}+a_{14}+a_{15}=90.$$

2) From the above we know $\{g_5\}:\{a_{23,}\,a_{24,}\,a_{25}\}$.

and
$$2T_3=T_1+T_5,$$

we have
$$2(a_{13}+a_{14}+a_{15})=(a_3+a_4+a_5)+(a_{23}+a_{24}+a_{25})$$
$$a_{23}+a_{24}+a_{25}=150.$$

3) Since a_4 and a_{44} are symmetrical about a_{24}, we need to compute a_4 and a_{24} first then can use the arithmetic mean of a_4 and a_{44} to get a_{44}. After solving two equation sets below,

and
$$\begin{cases} a_3+a_4+a_5=30 \\ 2a_4=a_3+a_5 \end{cases}$$
$$\begin{cases} a_{23}+a_{24}+a_{25}=150 \\ 2a_{24}=a_{23}+a_{25} \end{cases}$$

, we obtain $a_4=10$ and $a_{24}=50$. Since a_{24} is the arithmetic mean of a_4 and a_{44}, we have

$$2a_{24}=a_4+a_{44}$$
$$a_{44}=90.$$

Example 2.6.13   *Sequence of the Sums of Term Groups of A.S.*

If the sequence $\{a_n\}$ is an arithmetic sequence, $a_3+a_4+a_5+a_6+a_7+a_8=87$ and $a_{22}+a_{23}=131$, find $a_{37}+a_{38}+a_{39}+a_{40}+a_{41}+a_{42} = ?$

Solution:

Construct the term groups of the sequence $\{a_n\}$ with $m = 6$ and $p = 11$. Let

$$\{g_1\}:\{a_3, a_4, a_5, a_6, a_7, a_8\},$$
$$\{g_2\}:\{a_{20}, a_{21}, a_{22}, a_{23}, a_{24}, a_{25}\},$$
$$\{g_3\}:\{a_{37}, a_{38}, a_{39}, a_{40}, a_{41}, a_{42}\}.$$

By Property 7, we have an arithmetic sequence

$$\{T_k\}:\{ T_1, T_2, T_3 \}.$$

Because the term group $\{g_2\}$ is also an arithmetic sequence, we have

$$2(a_{22}+a_{23})=(a_{20}+a_{21})+(a_{24}+a_{25}).$$

Since $a_{22}+a_{23}=131$, then

$$T_2=a_{20}+a_{21}+a_{22}+a_{23}+a_{24}+a_{25}=393.$$

We also have

$$T_1=a_3+a_4+a_5+a_6+a_7+a_8=87$$

and

$$T_3=a_{37}+a_{38}+a_{39}+a_{40}+a_{41}+a_{42}$$

As $2T_2=T_1+T_3$,

$$T_3=2 \cdot T_2 - T_1 = 2 \cdot 393 - 87 = 699.$$

Therefore

$$a_{37}+a_{38}+a_{39}+a_{40}+a_{41}+a_{42}=699.$$

Example 2.6.14   *Sequence of the Sums of Term Groups of A.S.*

If the sequence $\{a_n\}$ is an arithmetic sequence, the sum of the first n terms $S_n = 32$, and the sum of the first $2n$ terms $S_{2n} = 128$. find $S_{5n} = ?$

Solution:

Build the term groups $\{G_k\}$ with $m = n$ and $p=0$. We have term groups like below

$$\{\{a_1, \cdots, a_n\}, \{a_{n+1}, \cdots, a_{2n}\}, \{a_{2n+1}, \cdots, a_{3n}\}, \{a_{3n+1}, \cdots, a_{4n}\}, \{a_{4n+1}, \cdots, a_{5n}\}, \cdots \}$$

$$\underbrace{\qquad}_{\{g_1\}} \quad \underbrace{\qquad}_{\{g_2\}} \quad \underbrace{\qquad}_{\{g_3\}} \quad \underbrace{\qquad}_{\{g_4\}} \quad \underbrace{\qquad}_{\{g_5\}} \quad \cdots$$

Then the sequence $\{T_k\}$ consisting of the sums of term groups of the sequence $\{a_n\}$ is an arithmetic sequence.

$$\{T_k\}:\{T_1, T_2, \cdots , T_k, \cdots\}$$

Since

$$T_1=S_n=a_1+\cdots+a_n,$$

$$T_2 = S_{2n} - S_n = a_{n+1} + \cdots + a_{2n},$$
$$T_3 = S_{3n} - S_{2n} = a_{2n+1} + \cdots + a_{3n},$$
$$\cdots,$$

we have
$$2T_3 = T_1 + T_5$$
$$S_{5n} = 2S_{4n} - S_{3n} + S_{2n} - 2S_n. \tag{a}$$

Since
$$2T_2 = T_1 + T_3$$
$$2(S_{2n} - S_n) = S_n + (S_{3n} - S_{2n})$$
$$S_{3n} = 3(S_{2n} - S_n) = 288.$$

Similarly
$$2T_3 = T_2 + T_4$$
$$S_{4n} = 3S_{3n} - 3S_{2n} + S_n$$

Substitute S_n, S_{2n}, S_{3n}, and S_{4n} into (a), we obtain
$$S_{5n} = 5(S_{3n} - S_{2n}) = 5(288 - 128) = 800.$$

▎ 2.7 Harmonic Sequences

We now discuss a new sequence called harmonic sequence. Actually it is a variation of an arithmetic sequence and we can discuss it in similar way.

DEFINITION   Harmonic Sequences

A sequence $\{a_n\}:\{a_1, a_2, a_3, \cdots, a_n, \cdots\}$ is a harmonic sequence if sequence

$$\frac{1}{a_1}, \frac{1}{a_2}, \frac{1}{a_3}, \cdots, \frac{1}{a_n}, \cdots$$

is an arithmetic sequence.

For instance, the nth term of harmonic sequences is the reciprocal of the nth term of arithmetic sequences. Two examples are given as below,

1) $1, \dfrac{1}{2}, \dfrac{1}{3}, \dfrac{1}{4}, \dfrac{1}{5}, \cdots, \dfrac{1}{n}, \cdots,$

2) $\dfrac{3}{2}, \dfrac{3}{4}, \dfrac{3}{6}, \dfrac{3}{8}, \cdots, \dfrac{3}{2n}, \cdots$

Example 2.7.1   *Harmonic Sequences*

If the sequence $\{a_n\}: \{\frac{3}{2}, \frac{3}{4}, \frac{3}{6}, \frac{3}{8}, \cdots\}$, determine if the sequence $\{a_n\}$ is a harmonic sequence and find the 20th term of $\{a_n\}$.

Solution:

Since $\{\frac{1}{a_n}\} = \{\frac{2}{3}, \frac{4}{3}, \cdots\}$ is an arithmetic sequence, the sequence $\{a_n\}$ is a harmonic sequence. We have the general term $a_n = \frac{3}{2n}$ and $a_{20} = \frac{3}{2 \cdot (20)} = \frac{3}{40}$. ▨

DEFINITION   Harmonic Mean

For any three terms a_{n-m}, a_n, and a_{n+m} of a harmonic sequence $\{a_n\}$, a_n is the **harmonic mean** of a_{n-m} and a_{n+m} and can be expressed as

$$2\frac{1}{a_n} = \frac{1}{a_{n-m}} + \frac{1}{a_{n+m}} \quad (n > 1, 0 < m < n) \quad (2.7.1)$$

Example 2.7.2   *Identifying Harmonic Sequences*

Prove that three numbers a, b, and c form a harmonic sequence if and only if $2\frac{1}{b} = \frac{1}{a} + \frac{1}{c}$.

Solution:

<u>Necessity</u>:

The sequence $\{\frac{1}{a}, \frac{1}{b}, \frac{1}{c}\}$ is an arithmetic sequence and we have $2\frac{1}{b} = \frac{1}{a} + \frac{1}{c}$.

<u>Sufficiency</u>:

As $2\frac{1}{b} = \frac{1}{a} + \frac{1}{c}$, $\frac{1}{a}, \frac{1}{b}, \frac{1}{c}$ is an arithmetic sequence. So the sequence $\{a, b, c\}$ is a harmonic sequence. ▨

▌ 2.8 Applications

Applications of arithmetic sequences can be seen in our real life frequently. The following examples show you these applications in various fields.

Example 2.8.1   *Contest Prize*

In a math contest, nine participants will win prizes. The first place wins $200, the second place $190, the third place $180, and so on. Every two adjacent prizes have the difference of $10. What is the prize for the last place?

Solution:

Because two adjacent prizes have the difference of $10, it is obvious that all prizes consist of an arithmetic sequence. Let the sequence $\{a_n\}$ be that arithmetic sequence, $a_1 = 200$, and $d = -10$.

$$a_1 = 200$$
$$a_2 = a_1 + (2-1)(-10)10 = 190$$
$$a_3 = a_1 + (3-1)(-10) = 180$$
$$\cdots$$
$$a_n = a_1 + (n-1)(-10) = 200 - 10(n-1) = 210 - 10n.$$

Because the last place is the 9th place, it is the 9th term of above sequence.

$$a_9 = 210 - 10 \cdot 9 = 120.$$

Then the last place wins $120.

Example 2.8.2   *Revenue of a Restaurant*

A restaurant's revenue is $80,000 in 2016. If its revenue increases by $10,000 each year in future, find the sum of the revenues for the period of 2016-2023.

Solution:

Since the revenue increases $10,000 every year it is a typical arithmetic sequence. Let the sequence $\{a_n\}$ be the arithmetic sequence, first term $a_1 = 80,000$ and the common difference $d = 10,000$. By the formula of arithmetic sequence, we obtain the revenue of the nth year.

$$a_n = 80,000 + (n-1) \cdot 10,000.$$

The sum of the revenues for the period of 2016-2023 will be the sum of the first 8 terms of the sequence $\{a_n\}$.

$$S_8 = 8 \cdot 80,000 + \frac{8(8-1) \cdot 10,000}{2} = 920,000.$$

Then the sum of the revenues for the period of 2016-2023 is $920,000.

64

Arithmetic Sequences

Example 2.8.3 *Payment by Installments*

David bought a TV of $1,000 and agreed to pay it by monthly installment after down payment $100. David will pay $100 each month plus 2% interest on the unpaid balance starting from next month.

1) How much will David pay for the 6th month?
2) How much will David pay for his TV in total when he pays it off.

Solution:

1) After paying the down payment of $100 on the date of purchase, there is the balance of $900 left for monthly installment starting from next month. Then it will be paid off in nine months.

The first month $\quad a_1 = 100 + 900 \cdot 0.02 = 118$

The second month $\quad a_2 = 100 + (900 - 100) \cdot 0.02 = 116$

... ...

The nth month $\quad a_n = 100 + (900 - 100(n-1)) \cdot 0.02 = 120 - 2n$

It is an arithmetic sequence, $a_1 = 118$, and $d = -2$. Let $\{a_n\}$ denote the sequence.

$$a_6 = 100 + (900 - 100(6-1)) \cdot 0.02 = 108$$

Thus David will pay $108 for the 6th month.

2) The the total payment by installments becomes

$$S_9 = 9 \cdot 118 + \frac{9 \cdot (9-1) \cdot (-2)}{2} = 990.$$

The total amount David paid for his TV is $100 + $990 = $1,090.

Example 2.8.4 *Payment by Installments*

Adam bought a car of $25,000 and agreed to pay back the loan by monthly installment after down payment $1,000. He will pay $400 plus the interest of 1% interest rate on the unpaid balance per month. List total monthly payments (principle + interest), interest and unpaid balance left for each month for the first six months.

Solution:

The balance before installments begins $24,000 (= $25,000 - $1,000). Adam's monthly installment of $400 is fixed then his all monthly payments consist of an arithmetic sequence, say $\{a_n\}$. Then the payment of the nth month will be

$$a_n = 400 + (24,000 - 400(n-1)) \cdot 0.01$$
$$= (400 + 240) + (n-1)(-4)$$

By Law 3 (1.2.12), this is an arithmetic sequence, $a_1 = 640$, and $d = -4$.

$$a_n = 400 + 240 + (n-1)(-4)$$

total monthly payment for the nth month   *principle paid*   *interest paid*

Month	1	2	3	4	5	6
Total Payment	$640	$636	$632	$628	$624	$620
Interest included	$240	$236	$232	$228	$224	$220
Unpaid balance	$23,600	$23,200	$22,800	$22,400	$22,000	$21,600

3   Geometric Sequences

▮ 3.1 Definition

Geometric sequences is an important type of sequence which the ratios of any two adjacent terms are the same. For example,

1) $3, 9, 27, 81, \cdots, 3^n, \cdots$ (ratio $= \dfrac{9}{3} = \dfrac{27}{9} = \dfrac{81}{27} = \cdots = \dfrac{3^n}{3^{n-1}} = 3$)

2) $-\dfrac{1}{2}, \dfrac{1}{4}, -\dfrac{1}{8}, \dfrac{1}{16}, \cdots$ (ratio $= \dfrac{1/4}{-1/2} = \dfrac{1/16}{-1/8} = \cdots = -\dfrac{1}{2}$)

3) $2, 1, 0.5, 0.25, 0.125, 0.0625, \cdots$ (ratio $= 1/2 = 0.5/1 = \cdots$)

DEFINITION **Geometric Sequences**

A sequence $\{a_n\}$ $(n \in N^+)$ is a **geometric sequence** if

$$\frac{a_{n+1}}{a_n} = r \quad\quad (\forall n, a_n \neq 0, r \neq 0)$$

r is a constant called **common ratio**.

Example 3.1.1   *The First Four Terms of a Geometric Sequence*

If a sequence $\{a_n\}$ is a geometric sequence, find the first four terms.
1) $r = 3, a_1 = 2$ 2) $r = 2, a_1 = -8$ 3) $r = -1/2, a_1 = 8$ 4) $r = -2, a_1 = 12$

Solution:

1) $\begin{cases} a_1 = 2 \\ a_2 = a_1 \cdot r = 2 \cdot 3 = 6 \\ a_3 = a_2 \cdot r = a_1 \cdot r^2 = 18 \\ a_4 = a_3 \cdot r = a_1 \cdot r^3 = 54 \end{cases}$ 2) $\begin{cases} a_1 = -8 \\ a_2 = a_1 \cdot r = (-8) \cdot 2 = -16 \\ a_3 = a_2 \cdot r = a_1 \cdot r^2 = -32 \\ a_4 = a_3 \cdot r = a_1 \cdot r^3 = -64 \end{cases}$

3) $\begin{cases} a_1 = 8 \\ a_2 = a_1 \cdot r = -4 \\ a_3 = a_2 \cdot r = a_1 \cdot r^2 = 2 \\ a_4 = a_3 \cdot r = a_1 \cdot r^3 = -1 \end{cases}$ 4) $\begin{cases} a_1 = 12 \\ a_2 = a_1 \cdot r = 12 \cdot (-2) = -24 \\ a_3 = a_2 \cdot r = a_1 \cdot r^2 = 48 \\ a_4 = a_3 \cdot r = a_1 \cdot r^3 = -96 \end{cases}$

▮ 3.2 General Term

By the definition of geometric sequence, the terms of a geometric sequence $\{a_n\}$ can be written in recursive form. Let a be a constant known.

$$\begin{cases} a_1 = a & (a \neq 0) \\ a_{n+1} = a_n r & (r \neq 0) \end{cases},$$

We can list these terms like

$$\begin{cases} a_1 = a \\ a_2 = a_1 r = a r \\ a_3 = a_2 r = a_1 r^2 = a r^2 \\ \cdots \\ a_n = a_{n-1} r = a_1 r^{n-1} = a r^{n-1} \\ \cdots \end{cases}$$

Now we can get the general term of a geometric sequence as below.

The General Term of a Geometric Sequence

Suppose $\{a_n\}$ is a geometric sequence then its general term is

$$a_n = a_1 \cdot r^{n-1} \qquad (n \in N^+, r \neq 0) \qquad (3.2.1)$$

When the subscript of the first term starts at m, the general term is

$$a_n = a_m \cdot r^{n-m} \qquad (m, n \in N^+, 1 \leqslant m < n) \qquad (3.2.2)$$

Example 3.2.1   *General Term of a Geometric Sequence*

Find the general term a_n of following geometric sequences.
1) $\{a_n\}: \{3, 6, 12, 24, 48, \cdots\}$ 2) $\{a_n\}: \{48, 24, 12, 6, 3, \cdots\}$
3) $\{a_n\}: \{3, -6, 12, -24, 48, \cdots\}$ 4) $\{a_n\}: \{1, 2, 2^2, 2^3, 2^4, \cdots\}$

Solution:

1) $a_1 = 3$ and $r = \dfrac{a_2}{a_1} = 2$, $a_n = a_1 \cdot r^{n-1} = 3 \cdot 2^{n-1}$.

2) $a_1 = 48$ and $r = \dfrac{a_2}{a_1} = \dfrac{24}{48} = \dfrac{1}{2}$, $a_n = 48 \cdot \left(\dfrac{1}{2}\right)^{n-1}$.

3) $a_1 = 3$ and $r = \dfrac{a_2}{a_1} = \dfrac{6}{3} = -2$, $a_n = 3 \cdot (-2)^{n-1}$.

4) $a_1 = 1$ and $r = \dfrac{a_2}{a_1} = 2$, $a_n = 2^{n-1}$.

Example 3.2.2   *General Term of a Geometric Sequence*

1) Insert two positive numbers between *1* and *125* to form a geometric sequence.
2) If a_1, a_2, and a_3 form a geometric sequence, $a_1 = 2$, and $a_1 + a_2 + a_3 = 26$, find a_n.

Solution:

1) Suppose the sequence $\{1, a_2, a_3, 125\}$ is a geometric sequence. Let a_2 and a_3 be two positive numbers to be inserted. As $a_4 = a_1 \cdot r^{4-1}$, $125 = 1 \cdot r^3$, then the common ratio $r = 5$.

$$a_2 = a_1 \cdot r^{2-1} = 1 \cdot 5^{2-1} = 5$$
$$a_3 = a_1 \cdot r^{3-1} = 1 \cdot 5^{3-1} = 25.$$

We obtain the geometric sequence $\{1, 5, 25, 125\}$.

2) Let the sequence $\{a_1, a_2, a_3\}$ be $\{2, 2r, 2r^2\}$,

then $a_1 + a_2 + a_3 = 2 + 2r + 2r^2 = 26.$
$$r^2 + r - 12 = 0$$

There are two real roots $r = 3$ and $r = -4$. Take $r = 3$ and the general term is
$$a_n = 2 \cdot 3^{n-1}.$$

Example 3.2.3   *General Term of a Geometric Sequence*

Suppose the sequence $\{a_n\}$ is a geometric sequence, if
1) $a_1 = 1$ and $a_n = a_1 + a_2 + \cdots + a_{n-1}$, find a_n
2) $a_4 = 27$ and $a_9 = 6561$, find a_n.
3) $a_3 = 8$, find $a_1 \cdot a_2 \cdot a_3 \cdot a_4 \cdot a_5 = ?$

Solution:

1) Since $a_n = a_1 + a_2 + \cdots + a_{n-2} + a_{n-1}$, $a_{n-1} = a_1 + a_2 + \cdots + a_{n-2}$.
$$a_n - a_{n-1} = a_{n-1}$$

As $\dfrac{a_n}{a_{n-1}} = 2$, we have $a_n = 2^{n-1}$.

2) Since $\{a_n\}$ is a geometric sequence,
$$\begin{cases} a_4 = a_1 \cdot r^3 = 27 \\ a_9 = a_1 \cdot r^8 = 6561 \end{cases}$$
$$\frac{a_9}{a_4} = \frac{r^8}{r^3} = r^5 = 243$$

Then $r = 3$ and $a_1 = 1$, The general term becomes $a_n = 3^{n-1}$.

3) Let r be the common ratio of $\{a_n\}$. We have $a_3 = a_1 \cdot r^2 = 8$ thus
$$a_1 \cdot a_2 \cdot a_3 \cdot a_4 \cdot a_5 = a_1 (a_1 \cdot r)(a_1 \cdot r^2)(a_1 \cdot r^3)(a_1 \cdot r^4)$$
$$= a_1^5 \cdot r^{10} = (a_1 \cdot r^2)^5 = 8^5 = 32768.$$

▌ 3.3 The Sum of the First *n* Terms

We already knew the sum of the first *n* terms of an infinite arithmetic sequence. Now we discuss the sum of the first *n* terms of an infinite geometric sequences. Suppose that a sequence $\{a_n\}$ is an infinite geometric sequence with common ratio *r*. We have

$$\{a_1, a_2, a_3, \cdots, a_n, \cdots\}$$

or
$$\{a_1, a_1 \cdot r, a_1 \cdot r^2, \cdots, a_1 \cdot r^{n-1}, \cdots\}.$$

By the equation (3.2.2), we have the sum of the first *n* terms of the sequence $\{a_n\}$ as below.

$$S_n = a_1 + a_2 + a_3 + \cdots + a_n = a_1 + a_1 \cdot r + a_1 \cdot r^2 + \cdots + a_1 \cdot r^{n-1} \quad \text{(a)}$$

Multiply *r* with two sides of equation (a). It, equivalently, shifts (a) one step to the right.

$$r \cdot S_n = a_1 \cdot r + a_1 \cdot r^2 + a_1 \cdot r^3 \cdots + a_1 \cdot r^n \quad \text{(b)}$$

(a) – (b):
$$(1-r) \cdot S_n = a_1 - a_1 \cdot r^n$$

When $r \neq 1$, we obtain
$$S_n = \frac{a_1(1-r^n)}{1-r}.$$

The Sum of the First *n* Terms of Geometric Sequences

The sum of the first *n* terms of an infinite geometric sequence $\{a_n\}$ is

$$S_n = \sum_{i=1}^{n} a_1 \cdot r^{n-1} = \begin{cases} na_1 & (r = 1) & (3.3.1) \\[2mm] \dfrac{a_1(1-r^n)}{1-r} & (r \neq 1) & (3.3.2) \\[2mm] \dfrac{a_1 - a_n r}{1-r} & (r \neq 1) & (3.3.3) \end{cases}$$

Let S_m be the sum of *m* consecutive terms starting at the term a_k of $\{a_n\}$, then

$$S_m = \frac{a_k - a_{k+m-1} r}{1-r} \quad (r \neq 1) \quad (3.3.4)$$

There are five basic quantities, a_1, a_n, *n* (or *m*), *r*, and S_n. We can obtain any two from other three.

Example 3.3.1   *Sum of the First n Terms of a Geometric Sequence*

Find the sum of the first several terms of a geometric sequence $\{a_n\}$ with common ratio r.

1) $a_1 = 64$, $a_6 = 10$, $r = -1/2$, $S_6 = ?$
2) $a_1 = 32$, $a_5 = 2$, $S_5 = ?$
3) $a_7 = 27$, $r = 3$, $S_7 = ?$

Solution:

1) $S_6 = \dfrac{a_1 - a_6 r}{1-r} = \dfrac{64 - 10 \cdot (-1/2)}{1-(-1/2)} = 46.$

2) $a_5 = a_1 \cdot r^4$ then $r = \dfrac{1}{2}$, $S_5 = \dfrac{a_1 - a_5 r}{1-r} = 62.$

3) Since $a_n = a_1 \cdot r^{n-1}$, $a_1 = \dfrac{a_n}{r^{n-1}} = \dfrac{a_7}{3^6} = \dfrac{1}{27}$, $S_7 = \dfrac{a_1 - a_7 r}{1-r} = \dfrac{1093}{27}.$

Example 3.3.2   *Sum of the First n Terms of a Geometric Sequence*

If $\{a_n\}$ is a geometric sequence, $a_8 - a_3 = 96$, $a_7 - a_4 = 24$, and $S_n = 255$, find n, a_1, and r.

Solution:

Let
$$\begin{cases} a_8 - a_6 = a_1 r^7 - a_1 r^5 = a_1 r^5(r^2-1) = 96 \\ a_6 - a_4 = a_1 r^5 - a_4 r^3 = a_1 r^3(r^2-1) = 24 \end{cases},$$

then $r = 2$. Substitute it into $a_1 r^3(r^2-1) = 24$ and $a_1 = 1$.

Let
$$S_n = \frac{a_1(1-r^n)}{1-r} = \frac{1-2^n}{1-2} = 2^n - 1 = 255,$$

then $n = 8$.

Example 3.3.3   *Sum of the First n Terms of a Geometric Sequence*

If $\{a_n\}$ is a geometric sequence, $S_3 = 21$, and $S_6 = 189$, find a_n and S_9.

Solution:

By the formula (3.3.2), $S_3 = \dfrac{a_1(1-r^3)}{1-r} = 21$ and $S_6 = \dfrac{a_1(1-r^6)}{1-r} = 189.$

$$\frac{S_6}{S_3} = \frac{a_1(1-r^6)}{a_1(1-r^3)} = \frac{189}{21} = 9$$

We have $1 + r^3 = 9$ and $r = 2$. Substitute it for r of the formula of S_3 or S_6,

$$\frac{a_1(1-2^3)}{1-2} = 21,$$

$a_1 = 3$, then
$$a_n = a_1 r^{n-1} = 3 \cdot 2^{n-1}$$
$$S_n = \frac{a_1(1-r^n)}{1-r} = 3 \cdot 2^n - 3.$$

Therefore
$$S_9 = 1533.$$

Example 3.3.4   *Sum of the First n Terms of a Geometric Sequence*

If two sequences $\{a_n\}$ and $\{b_n\}$ are given as below

- $a_1 = 1$ and $a_{n+1} = \frac{2n-1}{3n} S_{n+1}$,

- $b_n = n \cdot S_n$ (S_n is the sum of the first n term of the sequence $\{a_n\}$)

prove the sequence $\{b_n\}$ is a geometric sequence and find the general term b_n.

Solution:

Because $a_{n+1} = S_{n+1} - S_n$, we can write

$$\frac{2n-1}{3n} S_{n+1} - S_{n+1} - S_n$$

as

$$3n(S_{n+1} - S_n) = (2n-1)S_{n+1}.$$
$$\frac{(n+1)S_{n+1}}{nS_n} = 3.$$

Then $b_1 = 1 \cdot S_1 = 1 \cdot a_1 = 1$ and $\frac{b_{n+1}}{b_n} = 3.$

Therefore the sequence $\{b_n\}$ is a geometric sequence, $b_1 = 1$, $r = 3$, and $b_n = 3^{n-1}$.

Example 3.3.5   *Sum of the First n Terms of a Geometric Sequence*

Find the sum S_n of the first n terms of $\{a_n\}$: $\{1, 0.1, 0.01, 0.001, \cdots\}$.

Solution:

Since $1 = 10^{1-1}$, $0.1 = 10^{1-2}$, $0.01 = 10^{1-3}$, $\cdots$, the general term $a_n = 10^{1-n}$.

We have
$$\frac{a_{n+1}}{a_n} = \frac{10^{1-(n+1)}}{10^{1-n}} = \frac{1}{10}.$$

By the definition (3.1.1), the sequence $\{a_n\}$ is a geometric sequence with $a_1 = 1$ and $r = \frac{1}{10}$.

Thus its general term
$$a_n = 1 \cdot \left(\frac{1}{10}\right)^{n-1}.$$

By the formula (3.3.2),
$$S_n = \frac{10 - 10^{1-n}}{9}.$$

Example 3.3.6   Sum of the First n Terms of a Geometric Sequence

Suppose that S_n is the sum of the first n terms of the sequence $\{a_n\}$, $a_1 = 6$, and the sequence $\{S_n\}$ is a geometric sequence with common ratio $r = 2$. Find the general term a_n.

Solution:

Although $\{S_n\}$ is a geometric sequence we can not assume that the sequence $\{a_n\}$ is also a geometric sequence. So we use general method (1.4.4) .

$$\begin{cases} a_1 = S_1 \\ a_n = S_n - S_{n-1} \end{cases},$$

Because $S_1 = a_1 = 6$, $S_n = S_1 \cdot r^{n-1} = 6 \cdot 2^{n-1}$

and $S_{n-1} = S_1 \cdot r^{n-2} = 6 \cdot 2^{n-2}$,

$$a_n = S_n - S_{n-1} = 3 \cdot 2^{n-1}.$$

Example 3.3.7   Sum of the First n Terms of a Geometric Sequence

If the sequence $\{a_n\}$ is geometric and $a_n = 2 + 2^2 + \cdots + 2^n$, find the sum S_n of the first n terms of the sequence $\{a_n\}$.

Solution:

Because $a_n = 2 + 2^2 + \cdots + 2^n$ is the sum of the first n terms of the geometric sequence $\{2, 2^2, \cdots, 2^n\}$, we have

$$a_n = 2 + 2^2 + \cdots + 2^n = \frac{2(1 - 2^n)}{1 - 2} = 2^{n+1} - 2$$

Then the sum S_n of the first n terms of the sequence $\{a_n\}$ becomes

$$S_n = \sum_{i=1}^{n} a_n = \sum_{i=1}^{n} 2^{n+1} - \sum_{i=1}^{n} 2$$
$$= (2^2 + 2^3 + \cdots + 2^{n+1}) - 2n$$
$$= \frac{2^2(1 - 2^n)}{1 - 2} - 2n = 2^{n+2} - 2^2 - 2n.$$

Example 3.3.8   Sum of the First n Terms of a Geometric Sequence

There are two sequences $\{a_n\}$ and $\{b_n\}$, $a_1 = 2$, $a_{n+1} - 2a_n = 0$, and $b_n = 3a_{n+1} + 4a_{n-1}$, find the sum B_n of the first n terms of the sequence $\{b_n\}$.

Solution:

Since $a_1 = 2$ and $\dfrac{a_{n+1}}{a_n} = 2$, the sequence $\{a_n\}$ is a geometric sequence with $a_1 = 2$ and common ratio $r = 2$, the general term is

$$a_n = a_1 \cdot r^{n-1} = 2 \cdot 2^{n-1} = 2^n.$$

We have $a_{n+1} = 2^{n+1}$ and $a_{n-1} = 2^{n-1}$.

$$b_n = 3 \cdot 2^{n+1} + 4 \cdot 2^{n-1} = 2^4 \cdot 2^{n-1}$$

Thus the sequence $\{b_n\}$ is a geometric sequence with $b_1 = 2^4$ and common ratio $r = 2$. The sum B_n of the first n terms of the sequence $\{b_n\}$ is

$$S_n = 2^4 \frac{(1-2^n)}{1-2} = 2^{n+2} - 2^4.$$

Example 3.3.9   *Sum of the First n Terms of a Geometric Sequence*

If S_n is the sum of the first n terms of a sequence $\{a_n\}$, $a_1 = 1$ and $a_{n+1} = 3 S_n$ $(n \subset N^+)$, find
   1) a_n
   2) The sum of the first n terms of the sequence $\{n a_n\}$

Solution:

1) Because $a_{n+1} = S_{n+1} - S_n$, $S_{n+1} - S_n = 3 S_n$.

   Thus $\qquad\qquad \dfrac{S_{n+1}}{S_n} = 4.$

   The sequence $\{S_n\}$ is a geometric sequence with the first term $S_1 = a_1 = 1$ and common ratio $r = 4$. The general term $S_n = 4^{n-1}$. From $a_{n+1} = 3 S_n$ we have

   $$a_n = 3 S_{n-1} = 3 \cdot 4^{n-2} \qquad (n \geqslant 2).$$

   Thus $\qquad\qquad a_n = \begin{cases} 1 & (n=1) \\ 3 \cdot 4^{n-2} & (n \geqslant 2) \end{cases}$

2) $\{n a_n\}: \{1, 2 \cdot 3 \cdot 4^0, 3 \cdot 3 \cdot 4, 4 \cdot 3 \cdot 4^2, \dots, n \cdot 3 \cdot 4^{n-2}\}$

   Let S_n be the sum of the first n terms of $\{n a_n\}$.

   $$S_n = 1 + (2 \cdot 3 \cdot 4^0) + (3 \cdot 3 \cdot 4) + (4 \cdot 3 \cdot 4^2) + (4 \cdot 3 \cdot 4^3) + \dots + (n \cdot 3 \cdot 4^{n-2})$$
   $$-4 S_n = -4 - (2 \cdot 3 \cdot 4) - (3 \cdot 3 \cdot 4^2) - (4 \cdot 3 \cdot 4^3) - \dots - (n \cdot 3 \cdot 4^{n-1})$$
   $$S_n - 4 S_n = 3 + 3 \cdot 4 + 3 \cdot 4^2 + 3 \cdot 4^3 + \dots + 3 \cdot 4^{n-2} - n \cdot 3 \cdot 4^{n-1}$$
   $$-3 S_n = 3 \cdot (1 + 4 + 4^2 + 4^3 + \dots + 4^{n-2}) - n \cdot 3 \cdot 4^{n-1}$$

   Since $\qquad 1 + 4 + 4^2 + 4^3 + \dots + 4^{n-2} = \dfrac{1 \cdot (1 - 4^{n-2})}{1 - 4} = \dfrac{1}{3} 4^{n-2} - \dfrac{1}{3},$

   $$-3 S_n = 4^{n-2} - 1 - n \cdot 3 \cdot 4^{n-1}.$$

   Then $\qquad\qquad S_n = n \cdot 4^{n-1} - \dfrac{4^{n-2}}{3} + \dfrac{1}{3}.$

▌3.4 Geometric Mean

At first we give the definition of geometric mean of two numbers then we extend it to two terms of geometric sequences.

Geometric Mean of Two Numbers

For any three numbers a, b, c, the number b is called the **geometric mean** of a and c if

$$b^2 = a \cdot c \qquad (a \cdot c > 0) \qquad (3.4.1)$$

Example 3.4.1   *Geometric Mean of Two Numbers*

Find the geometric mean of two numbers.
   1) 6 and 24 2) $(x + 1)^2$ and $(x - 1)^2$

Solution:

1) $b^2 = a \cdot c = 6 \cdot 24 = 144$, $b = 12$.

2) $b^2 = a \cdot c = (x+1)^2 \cdot (x-1)^2$, $b = (x+1) \cdot (x-1) = x^2 - 1$.

Above concept can be applied to two terms of a geometric sequence $\{a_n\}$. Let the mid-term b represent any term, taking term a_n, then a and c will represent the two terms a_{n-m} and a_{n+m} $(m \geqslant 1)$, which both terms are symmetric about the term a_n.

$$a_1, a_2, a_3, \cdots, \underbrace{a_{n-m}, \cdots,}_{\substack{a \\ m}} \underbrace{a_{n-1}, a_n, a_{n+1}, \cdots, a_{n+m},}_{\substack{b \qquad m}} \cdots$$

Since $a_{n-m} = a_1 \cdot r^{n-m-1}$ and $a_{n+m} = a_1 \cdot r^{n+m-1}$,

$$a_{n-m} \cdot a_{n+m} = (a_1 \cdot r^{n-1})^2 = a_n^2.$$

Geometric Mean of Two Terms of a Geometric Sequence

The term a_n of a geometric sequence $\{a_n\}$ is the **geometric mean** of two terms a_{n-m} and a_{n+m}. It can be expressed as

$$a_n^2 = a_{n-m} \cdot a_{n+m} \quad (a_{n-m} \cdot a_{n+m} > 0, m \geqslant 1) \qquad (3.4.2)$$

Example 3.4.2   *Geometric Mean of Two Terms of G.S.*

Suppose the sequence $\{a_n\}$ is a geometric sequence. Determine if the following equations are true.

1) $a_7^2 = a_2 \cdot a_9$

2) $a_{18}^2 = a_{14} \cdot a_{22}$

Solution:

Check the symmetry of the subscript of two terms a_{n-m} and a_{n+m} about a_n.

1) **False**. $|7-2| \neq |7-9|$, the subscripts *2* and *9* are not symmetric about the subscript *7*.

2) **True**. $|18-14| = |18-22|$, the subscripts *14* and *22* are symmetric about the subscript *18*.

Example 3.4.3   *Geometric Mean of Two Terms of G. S.*

Three numbers *a*, *b*, and *c* form a geometric sequence, $a+b+c=28$, and $a \cdot b \cdot c = 512$, find these three numbers.

Solution:

Since the sequence $\{a, b, c\}$ is a geometric sequence, we let it be $\{\dfrac{b}{r}, b, b \cdot r.\}$

$$\begin{cases} \dfrac{b}{r}+b+b \cdot r = 28 \\ (\dfrac{b}{r}) \cdot b \cdot (b \cdot r) = 512 \end{cases} \implies \begin{cases} b(\dfrac{1}{r}+1+r)=28 & (a) \\ b^3 = 512 & (b) \end{cases}$$

From (b), we obtain $b = 8$. Substitute it for b of (a), we have a quadratic formula:

$$8r^2 - 20r + 8 = 0$$

Solve it and we get two real solutions

$$r_{1,2} = \frac{20 \pm \sqrt{20^2 - 4 \cdot 8 \cdot 8}}{2 \cdot 8} = \frac{5 \pm 3}{4},$$

then $r = 2$ and $r = 1/2$. We obtain these three numbers: 4, 8, 16.

Example 3.4.4   *Geometric Mean of Two Terms of G. S.*

If the sequence $\{a_n\}$ is a geometric sequence, $a_3 - 36$, $a_9 = \dfrac{4}{81}$, find a_6 and a_7.

Solution:

* Find a_6: Because the subscript 3 and 9 are symmetric about the subscript 6,

$$a_6^2 = a_3 \cdot a_9 \text{ and } a_6 = \frac{4}{3}.$$

* Find a_7: Use the formula (1.3.3) $a_n = a_m \cdot r^{n-m}$.

Let $a_9 = a_3 r^{9-3} = 36 \cdot r^6 = \dfrac{4}{81}$, then $r = \dfrac{1}{3}$.

Thus
$$a_7 = a_3 r^{7-3} = 36 \cdot \left(\dfrac{1}{3}\right)^4 = \dfrac{4}{9}.$$

Example 3.4.5   *Geometric Mean of Two Terms of G.S.*

If the sequence $\{3, x, y, 24, z\}$ is a geometric sequence, find x, y, and z.

Solution:

By the geometric mean, we have following four equations.

$$\begin{cases} x^2 = 3 \cdot y & \text{(a)} \\ y^2 = x \cdot 24 & \text{(b)} \\ 24^2 = y \cdot z & \text{(c)} \\ y^2 = 3 \cdot z & \text{(d)} \end{cases}$$

From (b) and (d) $z = 8\,x.$ (e)

Substitute (e) for z in (c), $24^2 = 8 \cdot x \cdot y$

$$y = \dfrac{24^2}{8 \cdot x} \qquad \text{(f)}$$

Substitute (f) for y in (a), we get $x = 6$. From (e) and (f) we obtain $y = 12$ and $z = 48$.

▌3.5 Identifying Geometric Sequences

To identify geometric sequences we can use the following laws.

The Laws of Identifying Geometric Sequences

The sequence $\{a_n\}$ is a geometric sequence if it meets any one of the following three necessary and sufficient conditions below.

  1) Law 1 (definition)

$$\frac{a_{n+1}}{a_n} = r \quad (\forall n \subset N^+, a_n \neq 0, \text{ constant } r \neq 0) \qquad (3.5.1)$$

  2) Law 2 (general term)

$$a_n = c \cdot q^n \quad (\forall n \in N^+, \text{ constants } c, q \neq 0) \qquad (3.5.2)$$

  3) Law 3 (geometric Mean)

$$a_n^2 = a_{n-m} \cdot a_{n+m} \quad (\forall n \in N^+, a_{n-m} \cdot a_{n+m} > 0, m \geqslant 1) \quad (3.5.3)$$

▌Proof   *Law 2 - (3.5.2)*

Necessity:

  For the sequence $\{a_n\}$ is a geometric sequence, $a_n = a_1 \cdot r^{n-1}$.

  Let $c = \left(\dfrac{a_1}{r}\right)$, $q = r$, then $a_n = c \cdot q^n$.

Sufficiency:

  Since $a_n = c \cdot q^n$ and $a_{n+1} = c \cdot q^{n+1}$, $\dfrac{a_{n+1}}{a_n} = q$. Thus $\{a_n\}$ is a geometric sequence.

Conclusion:

  The sequence $\{a_n\}$ is a geometric sequence $\Longleftrightarrow a_n = c \cdot q^n$.

▌Proof   *Law 3 - (3.5.3)*

Necessity:

  If the sequence $\{a_n\}$ is a geometric sequence, $a_{n-m} = a_1 \cdot r^{n-m-1}$ and $a_{n+m} = a_1 \cdot r^{n+m-1}$, then

$$a_{n-m} \cdot a_{n+m} = (a_1 \cdot r^{n-m-1}) \cdot (a_1 \cdot r^{n+m-1}) = (a_1 \cdot r^{n-1})^2 = a_n^2.$$

Sufficiency:

Since $a_{n+1}^2 = a_{n+1-m} \cdot a_{n+1+m}$,

$$\left(\frac{a_{n+1}}{a_n}\right)^2 = \frac{a_{n+1-m} \cdot a_{n+1+m}}{a_{n-m} \cdot a_{n+m}} = r^2.$$

For $\dfrac{a_{n+1}}{a_n} = r$ is a constant, the sequence $\{a_n\}$ is a geometric sequence.

Conclusion:

The sequence $\{a_n\}$ is a geometric sequence $\Longleftrightarrow a_n^2 = a_{n-m} \cdot a_{n+m}$.

Example 3.5.1   *Identifying Geometric Sequences*

Suppose $\{a_n\}$ is a geometric sequence with common ratio r, determine whether $\{b_m\}$ is a geometric sequence if

1) The sequence $\{b_m\}$ consists of all remained terms of $\{a_n\}$ by removing the first q terms.

2) The sequence $\{b_m\}$ consists of all terms with odd index n from $\{a_n\}$.

Solution:

1) Let

$$\{b_m\} = \{a_{q+1}, a_{q+2}, \cdots, a_{q+m-1}, a_{q+m}, \cdots\},$$

we have

$$\begin{cases} b_m = a_{q+m} = a_1 \cdot r^{q+m-1} \\ b_{m+1} = a_{q+m+1} = a_1 \cdot r^{q+m} \end{cases}$$

As

$$\frac{b_{m+1}}{b_m} = \frac{a_1 \cdot r^{q+m}}{a_1 \cdot r^{q+m-1}} = r$$

is a constant, the sequence $\{b_m\}$ is a geometric sequence.

2) Let

$$\{b_m\} = \{a_1, a_3, a_5, \cdots, a_{2n-1}, a_{2n+1}, \cdots\},$$

As

$$\frac{b_{m+1}}{b_m} = \frac{a_1 \cdot r^{2n}}{a_1 \cdot r^{2n-2}} = r^2$$

is a constant, the sequence $\{b_m\}$ is a geometric sequence.

Example 3.5.2   *Identifying Geometric Sequences*

1) If the sequence $\{a_n\}$ is a geometric sequence with common ratio r, determine if the sequence $\{a_n + q\}$ $(q \neq 0)$ is a geometric sequence.

2) The sequences $\{a_n\}$ and $\{b_n\}$ are geometric sequences, determine if the sequence $\{a_n \cdot b_n\}$ is a geometric sequence.

Solution:

1) $\{b_n\} = \{a_1 + q, a_2 + q, \cdots, a_n + q, \cdots\}$

$$\begin{cases} b_{n-1}=a_1 \cdot r^{n-2}+q \\ b_n=a_1 \cdot r^{n-1}+q \\ b_{n+1}=a_1 \cdot r^n+q \end{cases}$$

$$b_{n-1} \cdot b_{n+1}=(a_1 \cdot r^{n-2}+q) \cdot (a_1 \cdot r^n+q)=(a_1 \cdot r^{n-1})^2+q \cdot a_1(r^{n-2}+r^n)+q^2 \quad \text{(a)}$$

$$b_n^2=(a_1 \cdot r^{n-1}+q)^2=(a_1 \cdot r^{n-1})^2+2qa_1 \cdot r^{n-1}+q^2 \quad \text{(b)}$$

Compare (a) with (b). Since $2r^{n-1} \neq (r^{n-2}+r^n)$, $b_{n-1} \cdot b_{n+1} \neq b_n^2$. Then $\{b_n\}$ is not a geometric sequence.

2) Suppose $\{a_n\}$ and $\{b_n\}$ have the common ratio r and t respectively.
   Let $\{c_n\} = \{a_n \cdot b_n\}$, then

$$\begin{cases} c_n=(a_1 \cdot r^{n-1}) \cdot (b_1 \cdot t^{n-1}) \\ c_{n+1}=(a_1 \cdot r^n) \cdot (b_1 \cdot t^n) \end{cases}$$

The ratio $\dfrac{c_{n+1}}{c_n}=r \cdot t$ is a constant, then the sequence $\{c_n\}$ is a geometric s equence.

Example 3.5.3   *Identifying Geometric Sequences*

Two sequences $\{a_n\}$ and $\{b_n\}$ are geometric sequences and the sequence $\{c_n\} = \{a_n+ b_n\}$.
1) If $\{a_n\}$ and $\{b_n\}$ have the same common ratio r, determine that $\{c_n\}$ is a geometric sequence.
2) If $\{a_n\}$ and $\{b_n\}$ have different common ratio r_a and r_b, $r_a \neq r_b$, prove that $\{c_n\}$ is not a geometric sequence.

Solution:

1) For $a_n=a_1 \cdot r^{n-1}$ and $b_n=b_1 \cdot r^{n-1}$, $c_n=a_n+b_n=a_1 \cdot r^{n-1}+b_1 \cdot r^{n-1}=(a_1+b_1)r^{n-1}$.
   By Law 2, the sequence $\{c_n\}$ is a geometric sequence.

2) Since $a_n=a_1 \cdot r_a^{n-1}$ and $b_n=b_1 \cdot r_b^{n-1}$, $c_n=a_n+b_n=a_1 \cdot r_a^{n-1}+b_1 \cdot r_b^{n-1}$.
   Let's look at the first three terms

$$\begin{cases} c_1=a_1+b_1 \\ c_2=a_2+b_2=a_1 r_a+b_1 r_b \cdot \\ c_3=a_3+b_3=a_1 r_a^2+b_1 r_h^2 \end{cases}$$

If we can prove $c_2^2 \neq c_1 \cdot c_3$, then the sequence $\{c_n\}$ is not a geometric sequence.

$$c_1 \cdot c_3=(a_1+b_1)(a_1 r_a^2+b_1 r_b^2) = a_1^2 r_a^2+a_1 b_1(r_a^2+r_b^2)+b_1^2 r_b^2 \quad \text{(a)}$$

$$c_2^2=(a_1 r_a+b_1 r_b)^2=a_1^2 r_a^2+2a_1 b_1 r_a r_b+b_1^2 r_b^2 = a_1^2 r_a^2+2a_1 b_1 r_a r_b+b_1^2 r_b^2 \quad \text{(b)}$$

Compare (a) with (b). Since $r_a \neq r_b$, $r_a^2+r_b^2>2r_a r_b$.

Because $c_2^2 \neq c_1 \cdot c_3$, the sequence $\{c_n\}$ is not a geometric sequence.

Example 3.5.4   *Identifying Geometric Sequences*

Prove that the sequence $\{a^2+b^2, ab+bc, b^2+c^2\}$ is a geometric sequence if the sequence $\{a, b, c\}$ is a geometric sequence.

Solution:

Since the sequence $\{a, b, c\}$ is a geometric sequence, $b^2=ac$.

$$(ab+bc)^2=a^2b^2+2ab^2c+b^2c^2$$
$$=a^2b^2+b^2b^2+a^2c^2+c^2$$
$$=(a^2+b^2)(b^2+c^2)$$

By the law 3, the sequence $\{a^2+b^2, ab+bc, b^2+c^2\}$ is a geometric sequence.

Example 3.5.5   *Identifying Geometric Sequences*

S_n is the sum of the first n terms of a sequence $\{a_n\}$ and $S_n=3-2a_n$, determine if the sequence $\{a_n\}$ is a geometric sequence.

Solution:

If $n=1$, $a_1=S_1=3-2a_1$ and $a_1=1$.

If $n\geqslant 2$,
$$a_n=S_n-S_{n-1}$$
$$=a_n(3-2a_n)-(3-2a_{n-1})$$
$$=-2a_n+2a_{n-1}$$
$$\frac{a_n}{a_{n-1}}=\frac{2}{3}.$$

The sequence $\{a_n\}$ is a geometric sequence with common ratio $r=\dfrac{2}{3}$ and $a_1=1$.

When $n=1$, $a_1=1$ and it also meets the equation $S_n=3-2a_n$ then we have
$$a_n=\left(\frac{2}{3}\right)^{n-1}.$$

Example 3.5.6   *Identifying Geometric Sequences*

Suppose that $S_n=3^n-1$ is the sum of the first n terms of the sequence $\{a_n\}$. Determine if the sequence $\{a_n\}$ is a geometric sequence.

Solution:

We give three solutions as below.

1) Law 1:   As $S_n=3^n-1$, we have $S_{n-1}=3^{n-1}-1$ and $a_1=S_1=3-1=2$. Then
$$a_n=S_n-S_{n-1}=(3^n-1)-(3^{n-1}-1)=2\cdot 3^{n-1}.$$

Since $\dfrac{a_n}{a_{n-1}}=3$ is a constant, the sequence $\{a_n\}$ is a geometric

sequence.

2) Law 2:   The general term is $a_n=2\cdot3^{n-1}$. Let $c=\dfrac{2}{3}$ and $q=3$, then $a_n=c\cdot q^n$.

According to Law 2, the sequence $\{a_n\}$ is a geometric sequence.

3) Law 3:   Take any three consecutive terms, $a_{n-1}=2\cdot3^{n-2}$, $a_n=2\cdot3^{n-1}$ and $a_{n+1}=2\cdot3^n$ from $\{a_n\}$. We have

$$a_{n-1}\cdot a_{n+1}=(2\cdot3^{n-2})\cdot(2\cdot3^n)=(2\cdot3^{n-1})^2=a_n^2$$

By Law 3, the sequence $\{a_n\}$ is a geometric sequence.

Example 3.5.7   Identifying Geometric Sequences

Suppose that the sequence $\{a_n\}$ is a geometric sequence with common ratio r and the sequence

$$\{b_n\}:\{(a_1+a_2+a_3),\ (a_4+a_5+a_6),\ (a_7+a_8+a_9),\ \cdots,\ (a_{3n-2}+a_{3n-1}+a_{3n}),\ \cdots\}.$$

Is the sequence $\{b_n\}$ a geometric sequence? (Use two methods)

Solution:

1) Law 1 (definition):

$$b_1=a_1+a_2+a_3=S_3=\frac{a_1(1-r^3)}{1-r}$$

$$\cdots$$

$$b_n=a_{3n-2}+a_{3n-1}+a_{3n}=S_{3n}-S_{3(n-1)}$$
$$=\frac{a_1(1-r^{3n})}{1-r}-\frac{a_1(1-r^{3(n-1)})}{1-r}=\frac{a_1 r^{3(n-1)}(1-r^3)}{1-r}$$

$$b_{n+1}=a_{3(n+1)-2}+a_{3(n+1)-1}+a_{3(n+1)}=S_{3(n+1)}-S_{3n}=\frac{a_1 r^{3n}(1-r^3)}{1-r}$$

$$\cdots$$

As

$$\frac{b_{n+1}}{b_n}=\frac{r^{3n}}{r^{3(n-1)}}=r^3$$

is a constant, the sequence $\{b_n\}$ is a geometric sequence.

2) Law 3 (geometric mean):
Take any three consecutive terms $b_{n-1},\ b_n,\ b_{n+1}$, we have

$$b_{n-1}=\frac{a_1 r^{3(n-2)}(1-r^3)}{1-r}$$

and

$$b_{n+1}=\frac{a_1 r^{3n}(1-r^3)}{1-r}.$$

As

$$b_{n-1}\cdot b_{n+1}=(\frac{a_1 r^{3(n-1)}(1-r^3)}{1-r})^2=b_n^2$$

is a constant, the sequence $\{b_n\}$ is a geometric sequence.

▌ 3.6 Properties of Geometric Sequences

▶ Property 1

Monotonicity of Geometric Sequences

Suppose the sequence $\{a_n\}$ is a geometric sequence with common ratio r.

If	and	$\{a_n\}$ is a (an)	the extreme term in $\{a_n\}$
$r<0$		oscillating sequence	
$0<r<1$	$a_1<0$	increasing sequence	a_1 is the minimum term.
$0<r<1$	$a_1>0$	decreasing sequence	a_1 is the maximum term.
$r=1$		constant sequence	
$r>1$	$a_1<0$	decreasing sequence	a_1 is the maximum term.
$r>1$	$a_1>0$	increasing sequence	a_1 is the minimum term.

Example 3.6.1   *Types of Geometric Sequences*

Discuss the monotonicity of the following geometric sequences.

1) $a_n=3\cdot(-2)^{n-1}$ 2) $a_n=16\cdot\left(\dfrac{1}{2}\right)^{n-1}$ 3) $a_n=\dfrac{1}{2}\cdot 2^{n-1}$

Solution:

1) Because $r=-2<0$ and $r=-2<0$, $\{a_n\}$ is not a monotone sequence and it is an oscillating sequence.

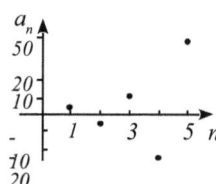

2) Because $0<r=\dfrac{1}{2}<1$ and $a_1=16>0$, $\{a_n\}$ is a decreasing sequence. $a_1=16$ is the maximum term in $\{a_n\}$.

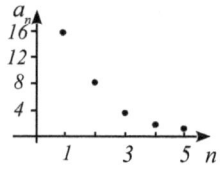

3) Because $r=2>0$ and $a_1=\dfrac{1}{2}>0$, $\{a_n\}$ is an increasing sequence. $a_1=\dfrac{1}{2}$ is the minimum term in $\{a_n\}$.

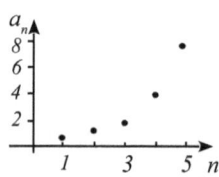

▶ Property 2

Suppose that $\{a_n\}$ is a geometric sequence with common ratio r and the constant $q \neq 0$.

1) If
$$b_n = q \cdot a_n, \qquad (3.6.1)$$
the sequence $\{b_n\}$ is a geometric sequence with common ratio r.

2) If
$$b_n = \frac{1}{a_n}, \qquad (3.6.2)$$
the sequence $\{b_n\}$ is a geometric sequence with common ratio $\frac{1}{r}$.

▌ Proof   *Property 2 - (3.6.1)*

Suppose that the sequence $\{a_n\}$ is a geometric sequence and the constant $q \neq 0$.

Let
$$b_n = q \cdot a_n,$$
and we have
$$b_{n+1} = q \cdot a_{n+1}.$$

$$\frac{b_{n+1}}{b_n} = \frac{q \cdot a_{n+1}}{q \cdot a_n} = \frac{a_{n+1}}{a_n} = r.$$

Because $\{a_n\}$ is a geometric sequence, $\dfrac{a_{n+1}}{a_n} = r$ is a constant. Then $\{b_n\}$ is a geometric sequence with $b_1 = q \cdot a_1$ and common ratio r.

▌ Proof   *Property 2 - (3.6.2)*

Suppose that the sequence $\{a_n\}$ is a geometric sequence.

Let
$$b_n = \frac{1}{a_n}.$$

As
$$b_{n+1} = \frac{1}{a_{n+1}},$$

$$\frac{b_{n+1}}{b_n} = \frac{a_n}{a_{n+1}} = \frac{1}{r}.$$

Because $\{a_n\}$ is a geometric sequence, r is a constant. Then the sequence $\{b_n\}$ is a geometric sequence with $b_1 = \dfrac{1}{a_1}$ and common ratio $\dfrac{1}{r}$.

Example 3.6.2   *Application of Property 2*

There are two sequences $\{a_n=4 \cdot 5^n\}$ and $\{b_n=\dfrac{3}{5^n}\}$, determine if the sequence

$\{b_n\}$ is a geometric sequence.

Solution:

By the law 2 of identifying geometric sequence, the sequence $\{a_n\}$ is a geometric sequence. Write b_n as

$$b_n=\frac{12}{4 \cdot 5^n}=12 \cdot \frac{1}{a_n}$$

By Property 2 (3.6.2), the sequence $\{\dfrac{1}{a_n}\}$ is a geometric sequence. Then the

sequence $\{b_n\}$ is a geometric sequence by the property 2, (3.6.1).

▶ Property 3

If two sequences $\{a_n\}$ and $\{b_n\}$ are geometric sequences with common ratio r and t respectively, then the sequence

$$\{a_n \cdot b_n\} \qquad\qquad (3.6.3)$$

is a geometric sequence with common ratio $r \cdot t$.

▌ Proof   *Property 3 - (3.6.3)*

Let $a_n=a_1 \cdot r^{n-1}$ and $b_n=b_1 \cdot t^{n-1}$.

$$a_n \cdot b_n=\left(a_1 \cdot r^{n-1}\right) \cdot \left(b_1 \cdot t^{n-1}\right)$$
$$a_{n+1} \cdot b_{n+1}=\left(a_1 \cdot r^n\right) \cdot \left(b_1 \cdot t^n\right)$$

$$\frac{a_{n+1} \cdot b_{n+1}}{a_n \cdot b_n}=\frac{\left(a_1 \cdot r^n\right) \cdot \left(b_1 \cdot t^n\right)}{\left(a_1 \cdot r^{n-1}\right) \cdot \left(b_1 \cdot t^{n-1}\right)}=r \cdot t$$

Because the common ratio (rt) is a constant, the sequence $\{a_n \cdot b_n\}$ is a geometric sequence.

▶ Property 4

If
  1) the sequence $\{a_n\}:\{a_1, a_2, \cdots, a_n, \cdots\}$ is a geometric sequence with common ratio r, and
  2) the sequence $\{b_m\}:\{b_1, b_2, \cdots, b_m, \cdots\}$ is an increasing arithmetic sequence of positive integers $(b_m \in N^+)$ with common difference d.
then the sequence

$$\{a_{b_1}, a_{b_2} \cdots, a_{b_m}, \cdots\} \qquad\qquad (3.6.4)$$

is a geometric sequence with common ratio r^d.

▌ Proof   Property 1 - (3.6.4)

Since $a_{b_{m+1}} = a_1 \cdot r^{b_{m+1}-1}$ and $a_{b_m} = a_1 \cdot r^{b_m-1}$,

we have

$$\frac{a_{b_{m+1}}}{a_{b_m}} = \frac{a_1 \cdot r^{b_{m+1}-1}}{a_1 \cdot r^{b_m-1}} = r^{b_{m+1}-b_m} = r^d.$$

Since the sequence $\{b_n\}$ is an arithmetic sequence, $b_{m+1} - b_m = d$ is a constant. Then common ratio r^d is a constant and the sequence $\{a_{b_1}, a_{b_2}, \cdots, a_{b_m}, \cdots\}$ is a geometric sequence with common ratio r^d.

Example 3.6.3   Application of Property 4

If the sequence $\{a_n\}:\{5 \cdot 2^n\}$ and the sequence $\{b_n\}:\{2n - 1\}$, determine if the sequence $\{c_n\}:\{a_{b_1}, a_{b_2} \cdots, a_{b_n}, \cdots\}$ is a geometric sequence. Find the sum of the first n terms of the sequence $\{c_n\}$.

Solution:

Determine if the sequence $\{a_n\}$ is a geometric sequence.

Since $a_n = 5 \cdot 2^n$ and $a_{n-1} = 5 \cdot 2^{n+1}$,

$$\frac{a_{n-1}}{a_n} = \frac{5 \cdot 2^{n+1}}{5 \cdot 2^n} = 2.$$

Then the sequence $\{a_n\}$ is a geometric sequence by Law 1.

Determine if the sequence $\{b_n\}$ is an increasing arithmetic sequence.

Since
$$b_{n+1} = 2(n+1) - 1 = 2n + 1,$$
$$d = b_{n+1} - b_n = 2 > 0,$$

the $\{b_n\}$ is an increasing sequence.

For $b_{n+1}-b_n=2$ is a constant, the sequence $\{b_n\}$ is an arithmetic sequence as well. Therefore, the sequence $\{c_n\}$ is a geometric sequence by Property 4.

Let C_n be the sum of the first n terms of the sequence $\{c_n\}$.
As $b_1=1, c_1=a_{b_1}=a_1=10$, and common ratio $r_c=r_a^d=2^2=4$,

$$C_n=\frac{10(1-4^n)}{1-4}=\frac{10}{3}(4^n-1).$$

Example 3.6.4   *Application of Property 4*

If $A_n=3^n-1$ is the sum of the first n terms of the sequence $\{a_n\}$ and $B_n=2\cdot n^2$ is the sum of the first n terms of the sequence $\{b_n\}$, determine if the sequence

$$\{a_{b_1}, a_{b_2}\cdots, a_{b_n},\cdots\}$$

is a geometric sequence.

Solution:

Determine if the sequence $\{a_n\}$ is a geometric sequence.

As $\qquad a_n=A_n-A_{n-1}=(3^n-1)-(3^{n-1}-1)=2\cdot3^{n-1},$

we have $\qquad\qquad a_{n-1}=2\cdot3^{n-2}.$

$$\frac{a_n}{a_{n-1}}=\frac{2\cdot3^{n-1}}{2\cdot3^{n-2}}=3$$

By Law 3, the sequence $\{a_n\}$ is a geometric sequence.

Determine if the sequence $\{b_n\}$ is an increasing arithmetic sequence.

Because the sun $B_n=2\cdot n^2$ is a quadratic function without constant term, the sequence $\{b_n\}$ is an arithmetic sequence by Law 4 of identifying arithmetic sequence (2.5.4).

The general term of $\{b_n\}$ is

$$b_n=B_n-B_{n-1}=2n^2-2(n-1)^2=4n-2$$
$$b_{n-1}=4(n-1)-2=4n-6$$
$$b_n-b_{n-1}=(4n-2)-(4n-6)=4>0$$

By the definition of increasing sequence, the sequence $\{b_n\}$ is an increasing sequence.

Therefore, by Property 4, the sequence $\{a_{b_1}, a_{b_2}\cdots, a_{b_n},\cdots\}$ is a geometric sequence.

▶ Property 5

Suppose $\{a_n\}$ is a geometric sequence with common ratio r.

If

1) The sequences $\{h_1, h_2, \cdots, h_m\}$ and $\{k_1, k_2, \cdots, k_m\}$ are two finite sequences of positive integers and both have m terms, and

2) $h_1 + h_2 + \cdots + h_m = k_1 + k_2 + \cdots + k_m$,

then

$$a_{h_1} \cdot a_{h_2} \cdot \ldots \cdot a_{h_m} = a_{k_1} \cdot a_{k_2} \cdot \ldots \cdot a_{k_m} \qquad (3.6.5)$$

Proof   Property 5 - (3.6.5)

Since the sequence $\{a_n\}$ is a geometric sequence we have

$$a_{h_1} \cdot a_{h_2} \cdot \ldots \cdot a_{h_m} = a_1^m \cdot r^{(h_1 + h_2 + \cdots + h_m - m)}$$

In the same way, $\qquad a_{k_1} \cdot a_{k_2} \cdot \ldots \cdot a_{k_m} = a_1^m \cdot r^{(k_1 + k_2 + \cdots + k_m - m)}$

Because $h_1 + h_2 + \cdots + h_m = k_1 + k_2 + \cdots + k_m$,

$$a_{h_1} \cdot a_{h_2} \cdot \ldots \cdot a_{h_m} = a_{k_1} \cdot a_{k_2} \cdot \ldots \cdot a_{k_m}.$$

Example 3.6.5   Application of Property 5

If the sequence $\{a_n\}$ is a geometric sequence and $a_5 \cdot a_6 = 10$, find $a_1 \cdot a_2 \cdot \ldots \cdot a_{10} = ?$

Solution:

By Property 5, we can use two methods to solve it.

1. Since $a_1 \cdot a_{10} = a_2 \cdot a_9 = a_3 \cdot a_8 = a_4 \cdot a_7 = a_5 \cdot a_6 = 10$,

$$a_1 \cdot a_2 \cdot \ldots \cdot a_{10} = (a_1 \cdot a_{10}) \cdot (a_2 \cdot a_9) \cdot (a_3 \cdot a_8) \cdot (a_4 \cdot a_7) \cdot (a_5 \cdot a_6)$$
$$= 100000$$

2. Since $a_1 \cdot a_2 \cdot a_9 \cdot a_{10} = a_3 \cdot a_4 \cdot a_7 \cdot a_8 = a_5 \cdot a_6 \cdot a_5 \cdot a_6 = 10 \cdot 10 = 100$,

$$a_1 \cdot a_2 \cdot \ldots \cdot a_{10} = (a_1 \cdot a_2 \cdot a_9 \cdot a_{10}) \cdot (a_3 \cdot a_4 \cdot a_7 \cdot a_8) \cdot (a_5 \cdot a_6)$$
$$= (100)(100)(10)$$

Example 3.6.6   Application of Property 5

Suppose that the sequence $\{a_n\}$ is a geometric sequence and $a_n > 0$.

1) If $a_2 u_5 u_8 = 64$ find $a_1 \cdot a_2 \cdot a_3 \ldots a_9 = ?$

2) If $a_1 \cdot a_2 \cdot a_9 + 3 a_3 \cdot a_5 \cdot a_7 + 3 a_4 \cdot a_6 \cdot a_8 + a_5 \cdot a_6 \cdot a_{10} = 216$, $a_4 + a_7 = ?$

Solution:

1) Since $a_5^2 = a_2 \cdot a_8$ and $a_2 a_5 a_8 = a_5^3 = 64$, $a_5 = 4$.

   As $a_5^2 = a_1 \cdot a_9 = a_2 a_8 = a_3 a_7 = a_4 a_6$,

$$a_1 a_2 a_3 a_4 a_5 a_6 a_7 a_8 a_9 = a_5^9 = 4^9.$$

2) As $a_1 \cdot a_2 \cdot a_9 = a_4^2$, $a_3 \cdot a_5 = a_4^2$, $a_6 \cdot a_8 = a_7^2$, and $a_5 \cdot a_6 \cdot a_{10} = a_7^3$, we have

$$a_1 \cdot a_2 \cdot a_9 + 3 \, a_3 \cdot a_5 \cdot a_7 + 3 \, a_4 \cdot a_6 \cdot a_8 + a_5 \cdot a_6 \cdot a_{10}$$
$$= a_4^3 + 3 \, a_4^2 \cdot a_7 + 3 \, a_4 \cdot a_7^2 + a_7^3$$
$$= (a_4 + a_7)^3 = 216.$$

Then $a_4 + a_7 = 6$.

▶ **Property 6**

Suppose
  1) The sequence $\{a_n\}$ is a geometric sequence with common ratio r.
  2) The term groups of $\{a_n\}$ are built by the rules in the section 1.1.7 and denoted by the following form.

$$\begin{cases} \cdots \\ \{g_{k-1}\}:\{a_{n-p-m}, \cdots, a_{n-p-1}\}, \\ \{g_k\}: \ \{a_n, \cdots, a_{n+m-1}\} \qquad (k, m, p \in N^+, 1 < m < n, p \geqslant 0) \\ \{g_{k+1}\}:\{a_{n+p+m}, \cdots, a_{n+p+2m-1}\} \\ \cdots \end{cases}$$

  3) $\{G_k\}:\{ \{g_1\}, \{g_2\}, \cdots, \{g_k\}, \cdots \}$.

Property 6.1
  If 1) T_k is the sum of all terms of the kth term group $\{g_k\}$, that is
$$T_k = a_n + a_{n+1} \cdots + a_{n+m-2} + a_{n+m-1} \text{ and}$$
  2) $\{T_k\}:\{T_1, T_2, \cdots, T_k, \cdots\}$

then the sequence $\{T_k\}$ is a geometric sequence with common ratio r^{m+p}.

Property 6.2
  If 1) P_k is the product of all terms of the kth term group $\{g_k\}$
$$P_k = a_n \cdot a_{n+1} \cdots a_{n+m-2} \cdot a_{n+m-1} \text{ and}$$
  2) $\{P_k\}$ denotes the sequence $\{P_1, P_2, \cdots, P_k, \cdots\}$

then the sequence $\{P_k\}$ is a geometric sequence with common ratio $r^{m(m+p)}$.

▌Proof   *Property 6.1*

Let the sequence $\{a_n\}$ be a geometric sequence with common ratio r. Construct term groups by the rules in the section 1.6.

$$\{G_k\}:\{\{g_1\}\ ,\ \cdots,\ \{g_k\},\ \{g_{k+1}\},\ \cdots\ \}.$$

Let the term group $\{g_k\}$ be the general term of the sequence $\{G_k\}$

$$\{g_k\}:\ \{a_n,\cdots,a_{n+m-1}\},$$

then
$$\{g_{k+1}\}:\{a_{n+p+m},\cdots,a_{n+p+2m-1}\}.$$

Let T_k be the sum of all terms of the kth term group $\{g_k\}$, we have the sequence

$$\{T_k\}:\{T_1,\ T_2,\ \cdots,\ T_k,\ \cdots\}.$$

Because the sequence $\{a_n\}$ is a geometric sequence and $a_n=a_1\cdot r^{n-1}$,

$$T_k=a_n+a_{n-1}+\cdots+a_{n+m-1}$$
$$-a_1\cdot r^{n-1}+a_1\cdot r^n+\cdots+a_1\cdot r^{n+m-2}$$
$$=a_1\cdot r^{n-1}\left(1+r+r^2+\cdots+r^{m-1}\right)$$

$$T_{k+1}=a_{n+p+m}+a_{n+p+m+1}+\cdots+a_{n+p+2m-1}$$
$$=a_1\cdot r^{n-p-m-1}+a_1\cdot r^{n-p-m}+\cdots+a_1\cdot r^{n-p-2}$$
$$=a_1\cdot r^{n+p+m-1}\left(1+r+r^2+\cdots+r^{m-1}\right).$$

Since the ratio $\dfrac{T_{k+1}}{T_k}=r^{m+p}$ is a constant the sequence $\{T_k\}$ is a geometric sequence with common ratio r^{m+p} by the definition of geometric sequences.

▌Proof   *Property 6.2*

Let the sequence $\{a_n\}$ be a geometric sequence. Build term groups of $\{a_n\}$ by the rules in the section 1.6.

$$\{G_k\}:\{\ \{g_1\}\ ,\ \cdots,\ \{g_k\},\ \{g_{k+1}\},\ \cdots\ \}.$$

Let term group $\{g_k\}$ be the general term of the sequence $\{G_k\}$

$$\{g_k\}:\{a_n,\cdots,a_{n+m-1}\},$$

then
$$\{g_{k+1}\}:\{a_{n+p+m},\cdots,a_{n+p+2m-1}\}.$$

Let P_k be the product of all terms of the kth term group $\{g_k\}$, we have the sequence

$$\{P_k\}:\{P_1, P_2, \cdots, P_k, \cdots\}$$

Also
$$P_k = a_n \cdot a_{n+1} \cdots a_{n+m-1}$$
$$= a_1 \cdot r^{n-1} \cdot a_1 \cdot r^n \cdots a_1 \cdot r^{n+m-2}$$
$$= a_1^m \cdot r^{(n-1)+n+\cdots+(n+m-2)}$$

Since sequence $\{(n-1), n, \cdots, (n+m-2)\}$, the power of r, is an arithmetic sequence that has m terms we can get the sum S_m by the formula (1.2.6).
$$S_m = \frac{m[(n-1)+(n+m-2)]}{2} = \frac{m(2n+m-3)}{2}.$$

Then
$$P_k = a_1^m \cdot r^{\frac{m(2n+m-3)}{2}}.$$

In the same way, let $P_{k+1} = a_{n+p+m} \cdot a_{n+p+m+1} \cdots a_{n+p+2m-1}$
$$= a_1 \cdot r^{n+p+m-1} \cdot a_1 \cdot r^{n+p+m} \cdots a_1 \cdot r^{n+p+2m-2}$$
$$= a_1^m \cdot r^{(n+p+m-1)+(n+p+m)+\cdots+(n+p+2m-2)}$$
$$= a_1^m \cdot r^{\frac{m(2n+2p+3m-3)}{2}}.$$

Since the ratio $\dfrac{P_{k+1}}{P_k} = r^{m(m+p)}$ is a constant, the sequence $\{P_k\}$ is a geometric sequence.

Example 3.6.7   *Application of Property 6.1*

Suppose that the sequence $\{a_n\}$ is a geometric sequence with common ratio r, S_n is the sum of the first n terms of $\{a_n\}$, and k is a positive integer. Prove the sequence
$$\{S_n, S_{2n} - S_n, S_{3n} - S_{2n}, \cdots, S_{kn} - S_{(k-1)n}, \cdots\}$$
is a geometric sequence.

Solution:

We construct term groups of the sequence $\{a_n\}$ with $m = n$ and $p = 0$.

$$\begin{cases} \{g_1\} : \{a_1, \cdots, a_n\} \\ \quad \cdots \\ \{g_k\} : \{a_{(k-1)n+1}, \cdots, a_{kn}\} \\ \quad \cdots \end{cases}$$

Let
$$\begin{cases} T_1 = S_n = a_1 + \cdots + a_n \\ T_2 = S_{2n} - S_n = a_{n+1} + \cdots + a_{2n} \\ \quad \cdots \\ T_k = S_{kn} - S_{(k-1)n} = a_{(k-1)n+1} + \cdots + a_{kn} \\ \quad \cdots \end{cases},$$

Let $$\{T_k\}:\{T_1, \cdots, T_k, \cdots\}.$$

By Property 6.1 of geometric sequences, the sequence $\{T_k\}$ is a geometric sequence.

Therefore the sequence
$$\{S_n, S_{2n} - S_n, S_{3n} - S_{2n}, \cdots, S_{kn} - S_{(k-1)n}, \cdots\}$$
is a geometric sequence.

Example 3.6.8   *Application of Property 6.2*

If the sequence $\{a_n\}$ is a geometric sequence, $a_1 \cdot a_2 = 8$, and $a_5 \cdot a_6 = 2048$, find $a_{13} \cdot a_{14} = ?$

Solution:

We construct term groups of the sequence $\{a_n\}$ with $m = 2$ and $p = 2$.

$$\begin{cases} \{g_1\}:\{a_1, a_2\}, \\ \{g_2\}:\{a_5, a_6\}, \\ \{g_3\}:\{a_9, a_{10}\}, \\ \{g_4\}:\{a_{13}, a_{14}\}. \end{cases}$$

Let
$$\begin{cases} P_1 = a_1 \cdot a_2 \\ P_2 = a_5 \cdot a_6 \\ P_3 = a_9 \cdot a_{10} \\ P_4 = a_{13} \cdot a_{14} \end{cases}$$

We have the sequence $\{P_k\}:\{P_1, P_2, P_3, P_4\}$.

By Property 6.2 of geometric sequences, the sequence $\{P_k\}$ is a geometric sequence and its common ratio is

$$r = \frac{P_2}{P_1} = \frac{2048}{8} = 256.$$

Then we obtain the general term P_k of $\{P_k\}$.

$$P_k = P_1 \cdot 256^{k-1}.$$

Therefore
$$a_{13} \cdot a_{14} = P_4 = 8 \cdot 256^{4-1} = 512^3.$$

3.7 Applications

In daily life, the knowledge of geometric sequences is mostly used in the fields related to the change of rate such as compound interest, loan, depression, annuity, population and so on.

Example 3.7.1   *Accumulated Distance*

A basketball falls freely at 10 feet high above floor. Every time the basketball hits the floor it renounces half the height of its fall. Find the total vertical distance the basketball went when it hits floor at the 10th times.

Solution:

The distance of the first fall is 10 feet.

The distance between the 1st hit and the 2nd hit is $2(10 \cdot 0.5)$ feet and let
$$a_1 = 20 \cdot 0.5 = 10.$$

The distance between the 2nd hit and the 3rd hit is $2(10 \cdot 0.5^2)$ feet and let
$$a_2 = 10 \cdot 0.5.$$

$$\cdots\cdots$$

The distance between the 9th hit and the 10th hit is $2(10 \cdot 0.5^9)$ feet and let
$$a_9 = 10 \cdot 0.5^8.$$

We have a geometric sequence whose $a_1 = 10$ and common ratio $r = 0.5$.
$$a_n = 10(0.5)^{n-1} \quad (1 \leqslant n \leqslant 9)$$
Let S_9 be the sum of all nine terms of $\{a_n\}$
$$S_9 = \frac{10(1-(0.5)^9)}{1-0.5} \approx 19.96 \text{ feet.}$$

Then the total distance the basketball traveled is $10 + 19.96 = 29.96$ feet.

▶ Compound Interest

Compound interest is the interest calculated on both the initial principle and the accumulated interest during prior periods.

Suppose that p is initial principle, r annual interest rate and m the number of compounding periods per year. Then the amount after n compounding periods becomes

$$a_n = p \cdot \left(1 + \frac{r}{m}\right)^n.$$

The total amount of interest accumulated after n compounding periods becomes

$$i_n = a_n - p.$$

Example 3.7.2   Compound Interest

David deposits $10,000 into his saving account at a interest rate 3% per year. List opening balance, interest and closing balance of the account for the first five months if the interest is compounded monthly.

Solution:

We have the following facts.
- Initial principle p = $10,000.
- Compound interest rate per year $r = 3\%$.
- Compounding periods per year $m = 12$.
- The number of compounding periods $n = 1, 2, 3, 4,$ and 5.

Then the amount after n compounding periods becomes

$$a_n = 10,000 \cdot \left(1 + \frac{0.03}{12}\right)^n.$$

Month	Opening balance	Accumulated interest	Closing balance a_n
1	$10,000	$25	$10,025
2	$10,025	$50.06	$10,050.06
3	$10,050.06	$75.19	$10,075.19
4	$10,075.19	$100.38	$10,100.38
5	$10,100.38	$125.63	$10,125.63

Example 3.7.3   Compound Interest

Mary invests $10,000 at 5% of interest rate per year. Find the amount after 10 years if the interest is compounded annually, quarterly, and monthly.

Solution:

We have the following facts
1) Initial principle p = $10,000.
2) Compound interest rate per year $r = 5\%$.

- **Annually**
  Compounding periods per year $m = 1$.
  Total compounding periods $n = 10$.

The amount after 10 years becomes

$$a_{10}=10,000\cdot\left(1+\frac{0.05}{1}\right)^{10}=\$\,16,288.95.$$

- **Quarterly**
  Compounding periods per year $m = 4$.
  Total compounding periods $n = 40$.
  The amount after 10 years becomes

$$a_{40}=10,000\cdot\left(1+\frac{0.05}{4}\right)^{40}=\$\,16,436.19.$$

- **Monthly**
  Compounding periods per year $m = 12$.
  Total compounding periods $n = 120$.
  The amount after 10 years becomes

$$a_{120}=10,000\cdot\left(1+\frac{0.05}{12}\right)^{120}=\$\,16,470.09.$$

▶ Annuities

Annuities are an investment products issued by an insurance company to help the seniors for their retirement. We discuss one of annuities, fixed annuity. It likes a term deposit investment which pays guaranteed rates of interest.

Let p be a fixed amount made at the beginning of every month during a period of time, r annual interest rate and m the number of compounding period per year. Then the balance after n compounding periods becomes

$$a_n = p\cdot\left(1+\frac{r}{m}\right)^1 + p\cdot\left(1+\frac{r}{m}\right)^2 + \cdots + p\cdot\left(1+\frac{r}{m}\right)^n.$$

On the right side of the equation, the first term is the balance of the nth (the last) deposit after gaining interest for 1 compounding period, the second term is the balance of the $(n-1)^{th}$ deposit after gaining interest for 2 compounding periods, ..., and the last term is the balance of the first deposit after gaining interest for n compounding periods. The total amount of interest accumulated after n compounding periods becomes

$$i_n = a_n - n\cdot p.$$

Example 3.7.4   *Annuities*

David makes deposit of $1,000 at the beginning of every month in his annuity investment account. If annual interest rate is 9% and compounded monthly. Find the balance of the account at the end of 10 years.

Solution:

The first deposit will gain interest for 120 months and its balance at the end of 10 years will be

$$a_{120}=1,000\cdot\left(1+\frac{0.09}{12}\right)^{120}=1,000\cdot(1.0075)^{120}.$$

The second deposit will gain interest for 119 months and its balance at the end of 10 years will be

$$a_{119}=1,000\cdot\left(1+\frac{0.09}{12}\right)^{119}-1,000\cdot(1.0075)^{119}.$$

$$\cdots\;\cdots$$

The last deposit will gain interest for 1 month and its balance at the end of 10 years will be

$$a_1=1,000\cdot\left(1+\frac{0.09}{12}\right)^{1}=1,000\cdot(1.0075)$$

The total balance becomes the sum of the first n terms of a geometric sequence.

$$1,000\cdot(1.0075)+1,000\cdot(1.0075)^2+\cdots+1,000\cdot(1.0075)^{120,}$$

$$S_{120}=1,000\cdot(1.0075\frac{1-(1.0075)^{120}}{1-1.0075})=194,965.63.$$

Then total balance at the end of 10 years will be $194,965.63.

4 Recursive Sequences

■ 4.1 Introduction

We learned that a sequences can be defined recursively in Chapter 1. For example,

1) $a_{n+1} = a_n + d$ (arithmetic sequences)
2) $a_{n+1} = a_n \cdot r$ (geometric sequences)
3) $a_{n+1} = (5n+3) \cdot a_n$
4) $a_{n+2} = 2a_{n+1} + 5a_n$

We notice that except for the first few initial terms, each term of a recursive sequence can be determined by its previous one or more terms and recursive relationship associated. Such sequences are called **recursive sequence**. The main task of solving a recursive sequence is to find its general term a_n from the recursive relationship given.

▶ Definition

To define a recursive sequence we should know the recursive relationship between the general term and some of its previous terms. Except for the first k terms of a recursive sequence any term a_n $(n > k)$ of the recursive sequence can be expressed as a function of its previous k terms,

$$a_n = f(a_{n-1}, a_{n-2}, \cdots, a_{n-k}).$$

DEFINITION   Recursive Sequences

A sequence $\{a_n\}$ is a **recursive sequence of order k** if
1) there are the first k given terms $a_1, a_2, \cdots, a_k$, called initial terms, and
2) the general term a_n is a function of its previous k terms

$$a_{n-1}, a_{n-2}, \cdots, a_{n-k}.$$

$$a_n = f(a_{n-1}, a_{n-2}, \cdots, a_{n-k}) \quad (1 \leqslant k < n \text{ and } n > k)$$

There are some examples below.

1) $a_{n+1} = 5a_n + 2n - 1$

2) $a_{n+2} = 2a_{n+1} + 5a_n$

3) $a_{n+2} = 2^n a_{n+1} + (3n+1)a_n$

4) $a_{n+3} = 3a_{n+2} - 2a_{n+1} + a_n + 1$

▶ Methods

We list some methods often used to find the general term of a recursive sequence.
Suppose $\{a_n\}$ is a recursive sequence.

1) Iterative addition

Arrange a_n $a_1 = (a_n - a_{n-1}) + (a_{n-1} - a_{n-2}) + \ldots + (a_2 - a_1)$.

If the right side of above equation is calculable, then

$$a_n = a_1 + (a_n - a_{n-1}) + (a_{n-1} \quad a_{n-2}) + \ldots + (a_2 - a_1).$$

2) Iterative product

Arrange $\quad \dfrac{a_n}{a_1} = \dfrac{a_n}{a_{n-1}} \cdot \dfrac{a_{n-1}}{a_{n-2}} \cdot \ldots \cdot \dfrac{a_2}{a_1} \quad (a_n \neq 0)$.

If $\dfrac{a_n}{a_{n-1}} \cdot \dfrac{a_{n-1}}{a_{n-2}} \cdot \ldots \cdot \dfrac{a_2}{a_1}$ is calculable,

$$a_n = a_1 \cdot \dfrac{a_n}{a_{n-1}} \cdot \dfrac{a_{n-1}}{a_{n-2}} \cdot \ldots \cdot \dfrac{a_2}{a_1}.$$

3) Substitution

Properly converting the format of a recursive equation by utilizing existing
identities, we can simplify the recursive equation. Two examples are given
as below.

(a) $a_{n+1} = p \cdot a_n + q \, a_{n-1} + c$

As $a_n = p \cdot a_{n-1} + q a_{n-2} + c$ then

$$a_{n+1} - a_n = p(a_n - a_{n-1}) + q(a_{n-1} - a_{n-2}).$$

Let $b_n - a_n - a_{n-1}$, we obtain $b_{n+1} = p b_n - q b_{n-1}$.

The recursive equation above becomes a new recursive equation

$$b_{n+1} = p b_n - q b_{n-1}.$$

(b) $a_{n+1} = p a_n + q^n$

Divide both sides by p^{n+1} and we have

$$\frac{a_{n+1}}{p^{n+1}}=\frac{a_n}{p^n}+\frac{q(n)}{p^{n+1}}.$$

Let

$$b_n=\frac{a_n}{p^n},$$

$$b_{n+1}=b_n+\frac{q(n)}{p^{n+1}}.$$

Let

$$f(n)=\frac{q(n)}{p^{n+1}},$$

we have

$$b_{n+1}=b_n+f(n).$$

The recursive equation above becomes a new recursive equation

$$b_{n+1}=b_n+f(n).$$

4) **Relationship between a_n and S_n**

When a_n and S_n are included in the formula of a recursive relationship, we can use the relationship between a_n and S_n,

$$a_n=\begin{cases} S_1 & (n=1) \\ S_n-S_{n-1} & (n\geqslant 2) \end{cases},$$

to convert the formula above into the one in terms of a_n and a_{n-1} (or S_n and S_{n-1}).

5) **Characteristic equation**

Suppose a recursive sequence is

$$a_n+p{\cdot}a_{n-1}+qa_{n-2}=0 \ (n>2, \ p,q{\in}R), \qquad (a)$$

It is called **recursive equation**. Let $a_n=x^n$, then we have the following equation

$$x^2+p{\cdot}x+q=0 \qquad\qquad (b)$$

We call the equation (b) the corresponding **characteristic equation** of the recursive equation (a). The characteristic equation (b) is a quadratic equation which may have two characteristic roots x_1 and x_2, decided by its discriminant $\Delta=p^2-4q$. The solution of the characteristic equation is the general term a_n of the recursive equation.

Example 4.1.1   *General Term of a Recursive Sequence*

1) If $a_1 = 2$ and $a_n - a_{n-1} = n^2$ $(n \geq 2)$, find a_5.

2) If $a_1 = 3$ and $\dfrac{a_n}{a_{n-1}} = 2$ $(n \geq 2)$, find a_5.

Solution:

1) List the first four equations

$$a_2 - a_1 = 2^2, a_3 - a_2 = 3^2, a_4 - a_3 = 4^2, a_5 - a_4 = 5^2.$$

Add these four equations

$$(a_2 - a_1) + (a_3 - a_2) + (a_4 - a_3) + (a_5 - a_4) = 2^2 + 3^2 + 4^2 + 5^2 = 54$$

We have $a_5 = a_1 + 54 = 2 + 54 = 56.$

2) List the first four equations

$$\frac{a_2}{a_1} = 2, \frac{a_3}{a_2} = 2, \frac{a_4}{a_3} = 2, \frac{a_5}{a_4} = 2.$$

Multiply these four equations $\dfrac{a_2}{a_1} \cdot \dfrac{a_3}{a_2} \cdot \dfrac{a_4}{a_3} \cdot \dfrac{a_5}{a_4} = 16$

Then $a_5 = a_1 \cdot 16 = 3 \cdot 16 = 48.$

Example 4.1.2   *General Term of a Recursive Sequence*

If $a_1 = 2$ and $a_{n+1} = 2 \cdot a_n + n$, find the general term a_n of the sequence $\{a_n\}$.

Solution:

Arrange $a_{n+1} = 2 \cdot a_n + n$ to

$$a_{n+1} + n = 2 \cdot (a_n + n),$$

and obtain the following $n - 1$ formulas:

$$\begin{cases} a_n + n = 2 \cdot (a_{n-1} + n) \\ a_{n-1} + n = 2 \cdot (a_{n-2} + n) \\ \dots \\ a_2 + n = 2 \cdot (a_1 + n) = 2 \cdot (2 + n) \end{cases}$$

By going backward from bottom to top, we have

$$\begin{cases} a_2 + n = 2^1 \cdot (2 + n) \\ a_3 + n = 2 \cdot (a_2 + n) = 2^2 \cdot (2 + n) \\ \dots \\ a_n + n = 2^{n-1} (2 + n) \end{cases}$$

From the last equation, we obtain the general term $a_n = 2^{n-1}(2 + n) - n.$

■ 4.2 Types of Recursive Sequences

Recursive sequences have different types and we will discuss the following types.

1. Type 1: $a_{n+1}=a_n+f(n)$
2. Type 2: $a_{n+1}=f(n)a_n$
3. Type 3: $a_{n+1}=pa_n+q$
4. Type 4: $a_{n+1}=pa_n+f(n)$
5. Type 5: $a_{n+2}=pa_{n+1}+qa_n$
6. Type 6: $a_{n+2}=pa_{n+1}+qa_n+c$

▶ Type 1: $a_{n+1} = a_n + f(n)$

Suppose $\{a_n\}$ is a recursive sequence,

$$a_{n+1}=a_n+f(n).$$

- If $f(n)$ is a constant, $a_{n+1}-a_n=d$, the sequence $\{a_n\}$ is an arithmetic sequence.

$$a_n=a_1+(n-1)d.$$

- If $f(n)$ is a function in term of n, we have

$$\begin{cases} a_n-a_{n-1}=f(n-1) \\ a_{n-1}-a_{n-2}=f(n-2) \quad (n\geqslant 2) \\ \dots \\ a_2-a_1=f(1) \end{cases}$$

Add all $n-1$ equations above

$$(a_n-a_{n-1})+(a_{n-1}-a_{n-2})+\cdots+(a_3-a_2)+(a_2-a_1)$$
$$= f(n-1)+f(n-2)+\cdots+f(2)+f(1)$$
$$a_n=a_1+\sum_{k=1}^{n-1} f(k) \qquad (n\geqslant 2)$$

If $\sum_{k=1}^{n-1} f(k)$ is calculable, we obtain the general term a_n of the sequence $\{a_n\}$.

The General Term of Recursive Sequences of Type 1

If a recursive sequence $\{a_n\}$ has the following recursive formula

$$a_{n+1}=a_n+f(n)$$

the general term a_n of $\{a_n\}$ can be expressed as

$$a_n=a_1+\sum_{k=1}^{n-1}f(k) \qquad (4.2.1)$$

Example 4.2.1   *Recursive Sequences of Type 1*

If $a_{n+1}=a_n+3n$, and $a_1=2$, find the general term a_n.

Solution:

Since $a_{n+1}-a_n=3n$, we have $\begin{cases} a_n-a_{n-1}=3(n-1) \\ a_{n-1}-a_{n-2}=3(n-2) \\ \cdots \\ a_2-a_1=3\cdot1 \end{cases}$ $(n\geqslant2)$

Add all $n-1$ equations above

$$(a_n-a_{n-1})+(a_{n-1}-a_{n-2})+\cdots+(a_2-a_1)=3(n-1)+3(n-2)+\cdots+3\cdot2+3\cdot1$$
$$a_n-a_1=3\cdot1+3\cdot2+\cdots+3(n-2)+3(n-1)$$
$$a_n=2+3(1+2+\cdots+(n-2)+(n-1))$$

As $\quad 1+2+\cdots+(n-2)+(n-1)=(n-1)\dfrac{(1+(n-1))}{2}=\dfrac{(n-1)\cdot n}{2},$

$$a_n=2+3\cdot S=2+3\cdot\frac{(n-1)\cdot n}{2}.$$

Example 4.2.2   *Recursive Sequences of Type 1*

Find the general term a_n of the following sequence if

1) $a_1=1,\ a_{n+1}=a_n+2^n$.

2) $a_1=1,\ \dfrac{n}{a_{n+1}}=\dfrac{(n+1)}{a_n}+2^n n(n+1)$.

3) $a_1=\dfrac{1}{2},\ a_{n+1}=a_n+\dfrac{1}{(n+1)(n+2)}$.

Solution:

1) Since $a_{n+1}-a_n=2^n$,

we have $\begin{cases} a_n-a_{n-1}=2^n \\ \cdots \\ a_2-a_1=2^1 \end{cases}$ $(n\geqslant2)$

Add all $n-1$ equations above

$$(a_n-a_{n-1})+(a_{n-1}-a_{n-2})+\cdots+(a_2-a_1)=2^{n-1}+2^{n-2}+\cdots+2^2+2^1$$
$$a_n-a_1=2^1+2^2+\cdots+2^{n-2}+2^{n-1}$$
$$a_n=1+2^1+2^2+\cdots+2^{n-2}+2^{n-1}$$

As the sequence $\{1,2,2^2.\ldots,2^{n-3},2^{n-2}\}$ is a geometric sequence,

$$a_n=1+2^1+2^2+\cdots+2^{n-2}+2^{n-1}=\frac{1-2^n}{1-2}=2^n-1$$

2) Arrange $\dfrac{n}{a_{n+1}}=\dfrac{(n+1)}{a_n}+2^n n(n+1)$ to

$$\frac{1}{(n+1)a_{n+1}}-\frac{1}{na_n}=2^n.$$

The sequence $\{\dfrac{1}{na_n}\}$ is a recursive sequence of Type 1. Use the result of the

question 1) above, we have $\dfrac{1}{na_n}=2^n-1.$

$$a_n=\frac{1}{n(2^n-1)}.$$

3) Since $a_{n+1}-a_n=\dfrac{1}{(n+1)(n+2)}$, we have $n-1$ equations.

$$\begin{cases} a_n-a_{n-1}=\dfrac{1}{n\cdot(n+1)} \quad (n\geqslant2) \\ \cdots \\ a_2-a_1=\dfrac{1}{2\cdot3} \end{cases}$$

Add all $n-1$ equations above

$$(a_n-a_{n-1})+(a_{n-1}-a_{n-2})+\cdots+(a_2-a_1)=\frac{1}{2\cdot3}+\frac{1}{3\cdot4}+\cdots+\frac{1}{(n-1)\cdot n}+\frac{1}{n\cdot(n+1)}$$
$$a_n-a_1=\frac{1}{2\cdot3}+\frac{1}{3\cdot4}+\cdots+\frac{1}{(n-1)\cdot n}+\frac{1}{n\cdot(n+1)}$$
$$a_n=\frac{1}{2}+(\frac{1}{2}-\frac{1}{3})+(\frac{1}{3}-\frac{1}{4})+\cdots+(\frac{1}{n-1}-\frac{1}{n})+(\frac{1}{n}-\frac{1}{n+1})$$
$$=\frac{1}{2}+(\frac{1}{2}-\frac{1}{n+1})=1-\frac{1}{n+1}.$$

Example 4.2.3   *Recursive Sequences of Type 1*

If $a_1 = 1$ and $n\,a_{n+1} = (n+1)\,a_n + 2\,n^3 + 3\,n^2 + n$, find the sum of the first n terms of

the sequence $\{\dfrac{a_n}{n^2}\}$.

Solution:

Arrange

$$n\,a_{n+1} = (n+1)\,a_n + 2\,n^3 + 3\,n^2 + n$$

to

$$a_{n+1} = \frac{n+1}{n}\,a_n + (2\,n+1)(n+1).$$

Divide both sides by $n+1$,

$$\frac{a_{n+1}}{n+1} = \frac{a_n}{n} + (2\,n+1).$$

Let $b_n = \dfrac{a_n}{n}$, then $b_1 = a_1 = 1$ and $b_{n+1} = b_n + (2\,n+1)$. It is a recursive sequence of

Type 1 and we have $n-1$ equations.

$$\begin{cases} b_n - b_{n-1} = (2n-1) \\ b_{n-1} - b_{n-2} = (2n-3) \\ \quad\cdots \\ b_2 - b_1 = (2\cdot1+1) \end{cases}$$

Add all these $n - 1$ equations.

$$(b_n - b_{n-1}) + (b_{n-1} - b_{n-2}) + \ldots + (b_2 - b_1) = (2n-1) + (2\,n-3) + \cdots + (2\cdot1+1)$$
$$b_n - b_1 = (2n-1) + (2n-3) + \cdots + (2\cdot2+1) + (2\cdot1+1)$$
$$b_n = 1 + 3 + 5 + 7 + \cdots + (2\,n-3) + (2\,n-1).$$

The right side of the equation above is the sum S_n of the arithmetic sequence $\{2\,n-1\}$ which has the first term $a_1 = 1$ and common difference $d = 2$.

By the formula (2.3.1), $\qquad\qquad S_n = n + \dfrac{n(n-1)\cdot2}{2} = n^2.$

$$b_n = S_n = n^2$$

Since $b_n = \dfrac{a_n}{n}$, $a_n = n^3$. $\qquad\qquad \dfrac{a_n}{n^2} = n$

The general term of the sequence $\{\dfrac{a_n}{n^2}\}$ is n. Then the sum S_n of the first n terms

of the sequence $\{\dfrac{a_n}{n^2}\}$ becomes

$$S_n = 1 + 2 + 3 + \cdots + n = \frac{n(n+1)}{2}.$$

▶ Type 2: $a_{n+1} = f(n) \cdot a_n$

Suppose $\{a_n\}$ is a recursive sequence and $a_{n+1} = f(n) \cdot a_n$.

- If $f(n)$ is a constant r, $\dfrac{a_{n+1}}{a_n} = r$, $\{a_n\}$ is a regular geometric sequence discussed in Chapter 3.

- If $f(n)$ is a function in terms of n, $\dfrac{a_{n+1}}{a_n} = f(n)$.

$$\frac{a_n}{a_{n-1}} = f(n-1), \frac{a_{n-1}}{a_{n-2}} = f(n-2), \cdots, \frac{a_2}{a_1} = f(1) \qquad (n \geq 2)$$

Multiply all $n-1$ equations above

$$a_n = \frac{a_n}{a_{n-1}} \cdot \frac{a_{n-1}}{a_{n-2}} \cdots \cdots \frac{a_2}{a_1} \cdot a_1 = f(n-1) \cdot f(n-2) \cdots f(1) \cdot a_1 = a_1 \cdot \prod_{k=1}^{n-1} f(k) \ (n \geq 2)$$

If $\displaystyle\prod_{k=1}^{n-1} f(k)$ is calculable, we can obtain the general term a_n.

The General Term of Recursive Sequences of Type 2

If a recursive sequence $\{a_n\}$ has the following recursive formula

$$a_{n+1} = f(n) \cdot a_n \qquad \text{($f(n)$ is not a constant)}$$

the general term a_n of the recursive sequence $\{a_n\}$ is

$$a_n = a_1 \cdot \prod_{k=1}^{n-1} f(k) \qquad (n \geq 2) \qquad\qquad (4.2.2)$$

Example 4.2.4   *Recursive Sequences of Type 2*

Find the general term a_n of the recursive sequence $\{a_n\}$ if $a_1 = 1$ and $a_{n+1} = 2^n a_n$.

Solution:

Since
$$\frac{a_{n+1}}{a_n} = 2^n,$$

we have $n-1$ equations

$$\frac{a_n}{a_{n-1}} = 2^{n-1}, \frac{a_{n-1}}{a_{n-2}} = 2^{n-2}, \cdots, \frac{a_3}{a_2} = 2^2, \frac{a_2}{a_1} = 2^1$$

Multiply all $n-1$ equations above

$$\frac{a_n}{a_{n-1}} \cdot \frac{a_{n-1}}{a_{n-2}} \cdots \cdots \frac{a_3}{a_2} \cdot \frac{a_2}{a_1} = 2^{n-1} \cdot 2^{n-2} \cdots \cdots 2^2 \cdot 2^1$$

$$a_n = a_1 \cdot 2^1 \cdot 2^2 \cdot \ldots \cdot 2^{n-2} \cdot 2^{n-1} = 2^{1+2+\ldots+(n-2)+(n-1)}$$

Then we obtain
$$a_n = 2^{\frac{n \cdot (n-1)}{2}}.$$

Example 4.2.5   Recursive Sequences of Type 2

Find the general term of the recursive sequence $\{a_n\}$ if $a_1 = 2$ and $a_n = \dfrac{n-1}{n+1} a_{n-1}$ $(n \geqslant 2)$.

Solution:

Since $a_1 = 2$, $a_{n-1} \neq 0$, we have

$$\frac{a_n}{a_{n-1}} = \frac{n-1}{n+1}, \quad \frac{a_{n-1}}{a_{n-2}} = \frac{n-2}{n}, \quad \frac{a_{n-2}}{a_{n-3}} = \frac{n-3}{n-1}, \quad \cdots, \quad \frac{a_3}{a_2} = \frac{2}{4}, \quad \frac{a_2}{a_1} = \frac{1}{3}$$

Multiply all $n - 1$ equations above

$$\frac{a_n}{a_{n-1}} \cdot \frac{a_{n-1}}{a_{n-2}} \cdot \frac{a_{n-2}}{a_{n-3}} \cdot \ldots \cdot \frac{a_4}{a_3} \cdot \frac{a_3}{a_2} \cdot \frac{a_2}{a_1} = \frac{n-1}{n+1} \cdot \frac{n-2}{n} \cdot \frac{n-3}{n-1} \cdots \frac{3}{5} \cdot \frac{2}{4} \cdot \frac{1}{3} = \frac{2 \cdot 1}{(n+1) \cdot n}$$

Then
$$a_n = a_1 \cdot \frac{2}{(n+1) \cdot n} = \frac{4}{(n+1) \cdot n}.$$

Example 4.2.6   Recursive Sequences of Type 2

Find the general term a_n of the recursive sequence $\{a_n\}$ if $a_1 = 2$ and $a_{n+1} = 2 n \cdot a_n - 2n + 1$.

Solution:

Arrange $a_{n+1} = 2 n \cdot a_n - 2n + 1$ to $a_{n+1} - 1 = 2n(a_n - 1)$. As $a_1 = 2$, $a_n - 1 \neq 0$, we have

$$\frac{a_{n+1} - 1}{a_n - 1} = 2n.$$

Then we list $n - 1$ equations

$$\frac{a_n - 1}{a_{n-1} - 1} = 2(n-1), \quad \frac{a_{n-1} - 1}{a_{n-2} - 1} = 2(n-2), \quad \cdots, \quad \frac{a_3 - 1}{a_2 - 1} = 2(2), \quad \frac{a_2 - 1}{a_1 - 1} = 2(1)$$

Multiply all $n - 1$ equations above

$$\frac{a_n - 1}{a_{n-1} - 1} \cdot \frac{a_{n-1} - 1}{a_{n-2} - 1} \cdot \frac{a_{n-2} - 1}{a_{n-3} - 1} \cdots \frac{a_3 - 1}{a_2 - 1} \cdot \frac{a_2 - 1}{a_1 - 1} = 2(n-1) \cdot 2(n-2) \cdot \ldots \cdot (2 \cdot 2) \cdot (2 \cdot 1)$$

$$a_n - 1 = (a_1 - 1) \cdot 2^{n-1} (n-1)! = 2^{n-1} (n-1)!,$$

we obtain

$$a_n = 2^{n-1} (n-1)! + 1.$$

Example 4.2.7   *Recursive Sequences of Type 2*

Find the general term a_n of the recursive sequence $\{a_n\}$ if $a_1=3$ and $a_{n+1}=3(n+2)\cdot2^n\cdot a_n$.

Solution:

Rewrite $a_{n+1}=3(n+2)\cdot2^n\cdot a_n$ to

$$\frac{a_{n+1}}{a_n}=3(n+2)\cdot2^n.$$

Then we list $n-1$ equations

$$\frac{a_n}{a_{n-1}}=3(n+1)2^{n-1}, \frac{a_{n-1}}{a_{n-2}}=3n2^{n-2}, \cdots, \frac{a_3}{a_2}=3(2+2)2^2, \frac{a_2}{a_1}=3(1+2)2^1$$

Multiply all $n-1$ equations above

$$\frac{a_n}{a_{n-1}}\cdot\frac{a_{n-1}}{a_{n-2}}\cdot\ldots\cdot\frac{a_3}{a_2}\cdot\frac{a_2}{a_1}=3(n+1)2^{n-1}\cdot3n2^{n-2}\cdot\ldots\cdot3(2+2)2^2\cdot3(1+2)2^1$$

$$\frac{a_n}{a_1}=3^{n-1}\cdot(n+1)n(n-1)\ldots4\cdot3\cdot2^{1+2+\ldots+(n-2)+(n-1)}$$

Because $(n+1)n(n-1)\cdots4\cdot3=(n+1)!-2!$ and $1+2+\cdots+(n-1)=n(n-1)/2$, we obtain

$$a_n=3^n\cdot((n+1)!-2)\cdot2^{\frac{n(n-1)}{2}}.$$

▶ **Type 3:** $a_{n+1} = p \cdot a_n + q$

Suppose a recursive sequence $\{a_n\}$ has the following recursive formula

$$a_{n+1} = p \cdot a_n + q \quad (p \neq 0, p \text{ and } q \text{ are constants}).$$

- If $p = 1$, $\{a_n\}$ is an arithmetic sequence.
- If $q = 0$, $\{a_n\}$ is a geometric sequence.
- If $p \neq 1$ and $q \neq 0$, $\{a_n\}$ is a linear recursive sequence.

We use the following two methods

1) Difference method

   Since $a_{n+1} = p \cdot a_n + q$ and $a_n = p \cdot a_{n-1} + q$,

   $$a_{n+1} - a_n = p(a_n - a_{n-1}).$$

   $$\frac{a_{n+1} - a_n}{a_n - a_{n-1}} = p$$

   The sequence $\{a_{n+1} - a_n\}$ is a geometric sequence whose first term is $a_2 - a_1$ and common ratio is p. Then the general term of the sequence $\{a_{n+1} - a_n\}$ becomes

   $$a_{n+1} - a_n = (a_2 - a_1) \cdot p^{n-1}. \tag{a}$$

   As $a_{n+1} = p \cdot a_n + q$ we substitute it for a_{n+1} of the equation above to obtain the general term a_n.

2) Formula method

   Since $a_{n+1} = p \cdot a_n + q$ and $a_2 = p \cdot a_1 + q$, we can substitute a_{n+1} and a_2 into the equation (a) above.

   $$(p-1)a_n + q = (p \cdot a_1 + q - a_1) \cdot p^{n-1}.$$

   Divide both sides by $p - 1$, we obtain the formula

   $$a_n = \left(a_1 + \frac{q}{p-1}\right) \cdot p^{n-1} - \frac{q}{p-1}.$$

The General Term of Recursive Sequences of Type 3

If a recursive sequence $\{a_n\}$ has the following recursive formula

$$a_{n+1} = p \cdot a_n + q \qquad (p \neq 0, p \text{ and } q \text{ are constants})$$

the general term a_n of $\{a_n\}$ is

$$a_n = \left(a_1 + \frac{q}{p-1}\right) p^{n-1} - \frac{q}{p-1} \qquad (n \geqslant 2) \tag{4.2.3}$$

Example 4.2.8   *Recursive Sequences of Type 3*

If $a_1 = 7$ and $a_{n+1} = 3a_n + 4$, find the general term a_n using two methods.

Solution:

1) Difference method

Since and $a_n = 3a_{n-1} + 4$, $a_{n+1} - a_n = 3(a_n - a_{n-1})$. We have

$$\frac{a_{n+1} - a_n}{a_n - a_{n-1}} = 3.$$

$\{a_{n+1} - a_n\}$ is a geometric sequence, its first term $a_2 - a_1 = (3a_1 + 4) - a_1 = 18$ and common ratio $r = 3$. The general term is

$$a_{n+1} - a_n = (a_2 - a_1)3^{n-1} = 2 \cdot 3^{n+1}.$$

Substitute $a_{n+1} = 3a_n + 4$ for a_{n+1} of the equation above,

$$a_n = 3^{n+1} - 2.$$

2) Formula method

Using the formula (4.2.3) with $a_1 = 7$, $p = 3$ and $q = 4$,

$$a_n = \left(7 + \frac{4}{(3-1)}\right)3^{n-1} - \frac{4}{(3-1)} = 3^{n+1} - 2.$$

Example 4.2.9   *Recursive Sequences of Type 3*

If $a_1 = 2$ and $a_{n+1} = \dfrac{2a_n}{4a_n + 1}$, find the general term a_n.

Solution:

Arrange $a_{n+1} = \dfrac{2a_n}{4a_n + 1}$ to $\dfrac{1}{a_{n+1}} = \dfrac{1}{2} \cdot \dfrac{1}{a_n} + 2$. Since $\dfrac{1}{a_n} = \dfrac{1}{2} \cdot \dfrac{1}{a_{n-1}} + 2$, we have

$$\frac{\dfrac{1}{a_{n+1}} - \dfrac{1}{a_n}}{\dfrac{1}{a_n} - \dfrac{1}{a_{n-1}}} = \frac{1}{2}.$$

Thus the sequence $\{\dfrac{1}{a_{n+1}} - \dfrac{1}{a_n}\}$ is a geometric sequence when $q = 0$. It has its first term $\dfrac{1}{a_2} - \dfrac{1}{a_1} = \dfrac{9}{4} - \dfrac{1}{2} = \dfrac{7}{4}$ and common ratio $r = \dfrac{1}{2}$. Thus $\dfrac{1}{a_{n+1}} - \dfrac{1}{a_n} = \dfrac{7}{4} \cdot \left(\dfrac{1}{2}\right)^{n-1}$.

Substitute $a_{n+1} = \dfrac{2a_n}{4a_n + 1}$ for a_{n+1} of the above, we obtain

$$a_n = \frac{2^n}{4 \cdot 2^n - 7}.$$

Example 4.2.10   *Recursive Sequences of Type 3*

If $a_1 = 2$ and $a_{n+1} = 2a_n + 1$, find the general term a_n using two methods.

Solution:

1) Difference method

Since $a_n = 2a_{n-1} + 1$, $a_{n+1} - a_n = 2(a_n - a_{n-1})$, we have

$$\frac{a_{n+1} - a_n}{a_n - a_{n-1}} = 2.$$

Thus $\{a_{n+1} - a_n\}$ is a geometric sequence whose first term $a_2 - a_1 = (2a_1 + 1) - a_1 = 3$ and common ratio $r = 2$. The general term

$$a_{n+1} - a_n = 3 \cdot 2^{n-1}.$$

Substitute $a_{n+1} = 2a_n + 1$ for a_{n+1} of the above we obtain

$$a_n = 3 \cdot 2^{n-1} - 1.$$

2) Formula method

As $a_{n+1} = 2a_n + 1$, take $a_1 = 2$, $p = 2$ and $q = 1$.

$$a_n = \left(2 + \frac{1}{(2-1)}\right)2^{n-1} - \frac{1}{(2-1)} = 3^{n+1} - 2 = 3 \cdot 2^{n-1} - 1.$$

Example 4.2.11   *Recursive Sequences of Type 3*

If S_n is the sum of the first n terms of the sequence $\{a_n\}$, $a_1 = 1$, and $3S_{n+1} + 2\dfrac{S_{n+1}}{S_n} = 1$ $(n \geq 2)$, find the general term a_n.

Solution:

Arrange $3S_{n+1} + 2\dfrac{S_{n+1}}{S_n} = 1$ to

$$\frac{1}{S_{n+1}} = 2\frac{1}{S_n} + 3.$$

The sequence $\{\dfrac{1}{S_n}\}$ is a recursive sequence of Type 3. Substitute $\dfrac{1}{S_1} = \dfrac{1}{a_1} = 1$, $p = 2$, and $q = 3$ into the formula (1.4.4)

$$\frac{1}{S_n} - \left(1 + \frac{3}{2-1}\right)2^{n-1} - \frac{3}{2-1} = 2^{n+1} \cdot 3$$

Then $S_n = \dfrac{1}{2^{n+1} - 3}$. Because $a_n = S_n - S_{n-1}$,

$$a_n = S_n - S_{n-1} = \frac{1}{2^{n+1} - 3} - \frac{1}{2^n - 3}.$$

Example 4.2.12   *Recursive Sequences of Type 3*

If $a_1 = 6$, $a_{n+1} \cdot a_n + 4 a_{n+1} - 4 a_n - 4 = 0$, and $b_n = \dfrac{4}{a_n - 2}$ find

    1) the general term b_n of the sequence $\{b_n\}$

    2) the sum of the first n term of the sequence $\{a_n \cdot b_n\}$.

Solution:

1) Arrange $b_n = \dfrac{4}{a_n - 2}$ to

$$a_n = \frac{4}{b_n} + 2$$

and substitute it into $a_{n+1} \cdot a_n + 4 a_{n+1} - 4 a_n - 4 = 0$. Then

$$\frac{2}{b_{n+1} \cdot b_n} + \frac{3}{b_{n+1}} - \frac{1}{b_n} = 0.$$

Multiply both sides by $b_{n+1} \cdot b_n$ and we have

$$b_{n+1} = 3 b_n + 2$$
$$b_{n+1} + 1 = 3(b_n + 1).$$
$$\frac{b_{n+1} + 1}{b_n + 1} = 3$$

Thus the sequence $\{b_n + 1\}$ is the geometric sequence with the first term

$$b_1 + 1 = \frac{4}{a_1 - 2} + 1 = 2$$

and common ratio $r = 3$. The general term of the sequence $\{b_n + 1\}$ can be written as

$$b_n + 1 = (b_1 + 1) \cdot r^{n-1} = 2 \cdot 3^{n-1}.$$

We obtain
$$b_n = 2 \cdot 3^{n-1} - 1.$$

2) Arrange $b_n = \dfrac{4}{a_n - 2}$ to

$$a_n \cdot b_n = 2 b_n + 4 = 4 \cdot 3^{n-1} + 2.$$

Since the sum of the first n terms of $\{a_n \cdot b_n\}$ equals that of the sequence $\{4 \cdot 3^{n-1} + 2\}$, let S_n be the sum of the first n terms of the sequence $\{4 \cdot 3^{n-1} + 2\}$.

$$S_n = \frac{4 \cdot (1 - 3^n)}{1 - 3} + 2n = 2 \cdot 3^n + 2n - 2.$$

▶ Type 4: $a_{n+1} = p \cdot a_n + f(n)$

Suppose that the recursive sequence $\{a_n\}$ has the following recursive formula

$$a_{n+1} = p \cdot a_n + f(n) \qquad (p \text{ is a constant and } p \neq 1)$$

A recursive sequence of Type 4 can be converted to a recursive sequence of Type 1.

1) Method 1

   Divide both sides of the equation above by p^{n+1}, we have

   $$\frac{a_{n+1}}{p^{n+1}} = \frac{a_n}{p^n} + \frac{f(n)}{p^{n+1}}$$

   Let $b_n = \dfrac{a_n}{p^n}$ and $f(n) = \dfrac{f(n)}{p^{n+1}}$,

   then $$b_{n+1} = b_n + f(n).$$

   Now the sequence $\{b_n\}$ is a recursive sequence of Type 1. If b_n is calculable we obtain

   $$a_n = b_n \cdot p^n.$$

2) Method 2

   If $f(n)$ is a linear function $f(n) = un + v$ ($u \neq 0$, u and v are constants), we can use the following method.

   Arrange $$a_{n+1} = p \cdot a_n + f(n) = p \cdot a_n + (un + v)$$
   $$a_n = p \cdot a_{n-1} + (u(n-1) + v) \qquad (n \geqslant 2)$$

   Then the difference between them is

   $$a_{n+1} - a_n = p \cdot (a_n - a_{n-1}) + (un(n-1) + v)$$
   $$= p(a_n - a_{n-1}) + v$$

   Let $b_n = a_n - a_{n-1}$, then $b_{n+1} = p b_n + v$.

   Now the sequence $\{b_n\}$ is a recursive sequence of Type 3. After finding $b_n = b(n)$ we obtain

   $$a_n = a_{n-1} + b(n).$$

   It is a recursive sequence of Type 1.

Example 4.2.13   Recursive Sequences of Type 4

Find the general term a_n of the sequence $\{a_n\}$ if $a_1 = 4$ and $a_{n+1} = 4a_n - 3n + 1$.

Solution:

Arrange $a_{n+1} = 4a_n - 3n + 1$ to $a_{n+1} - (n+1) = 4(a_n - n)$.

Let $b_n = a_n - n$, we have $\dfrac{b_{n+1}}{b_n} = 4$.

The sequence $\{b_n\}$ is the geometric sequence with the first term $b_1 = a_1 - 1 = 3$ and common ratio 4.

$$b_n = 3 \cdot 4^{n-1}.$$

Since $b_n = a_n - n$, we obtain $a_n = b_n + n = 3 \cdot 4^{n-1} + n$.

Example 4.2.14   Recursive Sequences of Type 4

Find the general term a_n if $a_1 = 2$ and $a_{n+1} = 2a_n + 6 \cdot (3^n)$.

Solution:

Divide both sides of the recursive equation $a_{n+1} = 2a_n + 6 \cdot (3^n)$ by 2^{n+1}.

$$\frac{a_{n+1}}{2^{n+1}} = \frac{a_n}{2^n} + \frac{3^{n+1}}{2^n}.$$

Let $b_n = \dfrac{a_n}{2^n}$, then $b_1 = \dfrac{a_1}{2} = 1$ and $b_{n+1} - b_n = \dfrac{3^{n+1}}{2^n}$.

From the above we can list $n - 1$ equations and add all of then.

$$b_n - b_{n-1} = \frac{3^n}{2^{n-1}}, \ b_{n-1} - b_{n-2} = \frac{3^{n-1}}{2^{n-2}}, \ \ldots, \ b_3 - b_2 = \frac{3^3}{2^2}, \ b_2 - b_1 = \frac{3^2}{2}$$

$$(b_n - b_{n-1}) + (b_{n-1} - b_{n-2}) + \cdots + (b_3 - b_2) + (b_2 - b_1)$$

$$= \frac{3^n}{2^{n-1}} + \frac{3^{n-1}}{2^{n-2}} + \cdots + \frac{3^3}{2^2} + \frac{3^2}{2}$$

Then

$$b_n - b_1 = 2\left[\left(\frac{3}{2}\right)^2 + \left(\frac{3}{2}\right)^3 + \cdots + \left(\frac{3}{2}\right)^{n-1} + \left(\frac{3}{2}\right)^n\right].$$

The right side of the equation above is the sum of all $n - 1$ terms of the geometric sequence with common ratio $r = \dfrac{3}{2}$. By the formula (3.3.4) the sum of $\{a_n\}$ is

$$S=\frac{3^{n+1}}{2^n}-\frac{3^2}{2}.$$

As

$$b_n=b_1+2\cdot S=\frac{3^{n+1}}{2^{n-1}}-8,$$

$$a_n=2^n b_n-2\cdot(3^n)-2^{n+3}.$$

Example 4.2.15   Recursive Sequences of Type 4

Find the general term a_n if $a_1=1$ and $a_{n+1}=4a_n+3n$.

Solution:

Since $a_{n+1}=4a_n+3n$, $a_n=4a_{n-1}+3(n-1)$. When $(n\geq2)$ we have the difference

$$a_{n+1}-a_n=4a_n+3n-4a_{n-1}-3(n-1)$$
$$=4(a_n-a_{n-1})+3$$

Let $b_n=a_{n+1}-a_n$ and substitute it for $a_{n+1}-a_n$ in above equation,

$$b_n=4b_{n-1}+3 \qquad\qquad (a)$$

The sequence $\{b_n\}$ is a recursive sequence of Type 3 and

$$b_1=a_2-a_1=(4a_1+3)-a_1=6$$

$$b_n=(6+\frac{3}{(4-1)})4^{n-1}-\frac{3}{(4-1)}=7\cdot(4^{n-1})-1$$

Then

$$a_{n+1}-a_n=7\cdot4^{n-1}-1 \qquad\qquad (b)$$

Since the equation (b) is a recursive sequence of Type 1, $a_{n+1}=a_n+f(n)$. From (b) we obtain $n-1$ equations

$$a_n-a_{n-1}=7\cdot4^{n-2}-1,\ a_{n-1}-a_{n-2}=7\cdot4^{n-3}-1,\ \dots\ ,\ a_2-a_1=7\cdot4^0-1.$$
$$(a_n-a_{n-1})+(a_{n-1}-a_{n-2})+\dots+(a_2-a_1)$$
$$=(7\cdot4^{n-2}-1)+(7\cdot4^{n-3}-1)+\dots+(7\cdot4^1-1)$$
$$a_n-a_1=(7\cdot4^{n-2}-1)+(7\cdot4^{n-3}-1)+\dots+(7\cdot4^2-1)+(7\cdot4^1-1)$$
$$a_n=1+7^{n-1}\cdot(4^1+4^2+\dots+4^{n-3}+4^{n-2})-(n-2)$$

Since the sequence $\{4^1+4^2+\dots+4^{n-3}+4^{n-2}\}$ is a geometric sequence whose the first term 4 and common ratio $r=4$. Then we have the sum S of the sequence.

$$S=4^1+4^2+\dots+4^{n-3}+4^{n-2}=\frac{4}{3}(4^{n-2}-1)$$

Then

$$a_n=1+7^{n-1}\cdot S-(n-2)=\frac{4}{3}7^{n-1}(4^{n-2}-1)-(n+1).$$

Example 4.2.16   *Recursive Sequences of Type 4*

If $S_n = 3a_n - 2 \cdot 3^{n+1}$ is the sum of the first n terms of the sequence $\{a_n\}$ find the general term a_n.

Solution:

From $S_n = 3a_n - 2 \cdot 3^{n+1}$ we have

$$S_{n+1} = 3a_{n+1} - 2 \cdot 3^{n+2}.$$

Since $a_{n+1} = S_{n+1} - S_n$,

$$a_{n+1} = 3 \cdot a_{n+1} - 2 \cdot 3^{n+2} - 3 \cdot a_n + 2 \cdot 3^{n+1}.$$

Then

$$a_{n+1} = \frac{3}{2} \cdot a_n - 2 \cdot 3^{n+1}. \qquad\qquad (a)$$

Now it is converted to a recursive sequence of Type 4.

Divide both sides by 3^{n+1},

$$\frac{a_{n+1}}{3^{n+1}} = \frac{3}{2} \cdot \frac{a_n}{3^{n+1}} - \frac{2 \cdot 3^{n+1}}{3^{n+1}}$$

$$\frac{a_{n+1}}{3^{n+1}} = \frac{1}{2} \cdot \frac{a_n}{3^n} - 2$$

Let $b_n = \dfrac{a_n}{3^n}$,

$$b_{n+1} = \frac{1}{2} \cdot b_n - 2.$$

It is a recursive sequence of Type 3.

As

$$b_n = \frac{1}{2} \cdot b_{n-1} - 2,$$

$$b_{n+1} - b_n = \frac{1}{2} \cdot (b_n - b_{n-1}).$$

The sequence $\{b_{n+1} - b_n\}$ is a geometric sequence. Since $a_1 = S_1 = 3a_1 - 2 \cdot 3^{1+1}$, $a_1 = 3^2$ and $a_2 = \dfrac{3}{2} \cdot a_1 - 2 \cdot 3^2 = \dfrac{-9}{2}$ from the formula (a). We have $b_1 = \dfrac{a_1}{3^1} = \dfrac{3^2}{3} = 3$, and $b_2 = \dfrac{a_2}{3^2} = -\dfrac{1}{2}$.

The sequence $\{b_{n+1} - b_n\}$ has its first term $b_2 - b_1 = -\dfrac{7}{2}$ and common ratio $r = \dfrac{1}{2}$.

$$b_{n+1} - b_n = -\frac{7}{2} \cdot \left(\frac{1}{2}\right)^{n-1} = -7 \cdot \frac{1}{2^n}$$

It is a recursive sequence of Type 1.

$$\begin{cases} b_n - b_{n-1} = -7 \cdot \dfrac{1}{2^{n-1}} \\ \cdots \\ b_2 - b_1 = -7 \cdot \dfrac{1}{2} \end{cases}$$

$$(b_n - b_{n-1}) + (b_{n-1} - b_{n-2}) + \cdots + (b_2 - b_1) = -7 \cdot \left(\frac{1}{2^{n-1}} + \frac{1}{2^{n-2}} + \cdots + \frac{1}{2} \right)$$

$$b_n - b_1 = -7 \cdot \left(1 - \frac{1}{2^{n-1}} \right)$$

$$b_n = -4 + 7 \cdot \frac{1}{2^{n-1}}$$

Therefore

$$a_n = 3^n \cdot b_n = 3^n \left(-4 + 7 \cdot \frac{1}{2^{n-1}} \right)$$

Example 4.2.17   *Recursive Sequences of Type 4*

If $a_{n+2} - 3a_{n+1} + 2a_n - 2^{n+1}$, $a_1 = 2$ and $a_2 - 8$ find a_n and S_n.

Solution:

Arrange

$$a_{n+2} - 3a_{n+1} + 2a_n = 2^{n+1}$$

to

$$(a_{n+2} - 2a_{n+1}) - (a_{n+1} - 2a_n) = 2^{n+1}.$$

Let $b_n = a_{n+1} - 2a_n$, then $b_{n+1} - b_n = 2^{n+1}$.

It is a recursive sequence of Type 3 and we have the following $n - 1$ equations.

$$\begin{cases} b_n - b_{n-1} = 2^n \\ b_{n-1} - b_{n-2} = 2^{n-1} \\ \cdots \\ b_3 - b_2 = 2^3 \\ b_2 - b_1 = 2^2 \end{cases}$$

Add all of them and we obtain

$$b_n - b_1 = 2^2 + 2^3 + 2^{4.} .. + 2^n = 2^{n+1} - 4.$$

Since $b_1 = a_2 - 2a_1 = 4$,

$$b_n = 2^{n+1}.$$

That is

$$a_{n+1} - 2a_n = 2^{n+1}.$$

Divide both sides by 2^{n+1}.

$$\frac{a_{n+1}}{2^{n+1}} - \frac{a_n}{2^n} = 1$$

Then the sequence $\{\frac{a_n}{2^n}\}$ is an arithmetic sequence. Its first term $\frac{a_1}{2}=1$ and common difference $d=1$.

$$\frac{a_n}{2^n}=\frac{a_1}{2}+(n-1)d=n.$$

Therefore

$$a_n=n\cdot2^n.$$

$$\begin{cases} S_1=a_1=2 & (n=1) \\ S_n=2^1+2\cdot2^2+3\cdot2^3+\cdots+n\cdot2^n(n\geqslant2) \end{cases} \qquad (a)$$

$$2S_n=2^2+2\cdot2^3+3\cdot2^4+\cdots+n\cdot2^{n+1} \qquad (b)$$

(a) – (b)

$$S_n-2S_n=2+(2^2+2^3+2^4+\cdots+2^n)-n\cdot2^{n+1}$$

Since

$$2^2+2^3+2^4+\cdots+2^n=\frac{2^2(1-2^{n-1})}{1-2}=2^{n+1}-4,$$

we obtain

$$S_n-2S_n=2+2^{n+1}-4-n\cdot2^{n+1}$$
$$-S_n=(1-n)2^{n+1}-2.$$

Then

$$S_n=(n-1)2^{n+1}+2.$$

▶ **Type 5:** $a_{n+2} = p \cdot a_{n+1} + q \cdot a_n$

In the introduction of this chapter characteristic equation method is introduced. Now we apply it to solve the recursive sequences of Type 5 like,

$$a_{n+2} = p a_{n+1} + q a_n.$$

The General Term of Recursive Sequences of Type 5

Suppose that a recursive sequence $\{a_n\}$ has the following recursive formula

$$a_{n+2} + p \cdot a_{n+1} + q a_n = 0 \quad (n \geqslant 1, q \neq 0, p \text{ and } q \text{ are real})$$

and two initial terms $a_1 = a, a_2 = b$. Then its corresponding characteristic equation is

$$x^2 + p \cdot x + q = 0. \tag{4.2.4}$$

The general term a_n of $\{a_n\}$ is decided by the discriminant of the quadratic equation

$$\Delta = p^2 - 4q.$$

- If $\Delta > 0$, there are two different real roots x_1 and x_2. Then

$$a_n = \lambda_1 x_1^n + \lambda_2 x_2^n \tag{4.2.5}$$

- If $\Delta = 0$, there are two same real roots $x_1 = x_2 = x_0$.

$$a_n = (\lambda_1 + n \cdot \lambda_2) x_0^n \tag{4.2.6}$$

(λ_1, λ_2 are real numbers to be decided by a_1 and a_2)

Example 4.2.18   *Recursive Sequences of Type 5*

Find the general term a_n of $\{a_n\}$ if $a_1 = \sqrt{2}$, $a_2 = 8$, and $a_{n+2} = \sqrt{8} a_{n+1} - 2 a_n \ (n \geqslant 1)$.

Solution:

The corresponding characteristic equation of the recursive formula

$$a_{n+2} = \sqrt{8} a_{n+1} - 2 a_n$$

is

$$x^2 - \sqrt{8} \cdot x + 2 = 0$$

Since the discriminant of the characteristic equation

$$\Delta = \left(-\sqrt{8}\right)^2 - 4 \cdot 2 - 0,$$

there are two same real roots $x_1 = x_2 = \sqrt{2}$. By equation (4.2.6), we have general term

$$a_n=(\lambda_1+\lambda_2\cdot n)\sqrt{2^n} \qquad \text{(a)}$$

Substitute $a_1=\sqrt{2}$, $a_2=8$ for a_n in above equation respectively, we have

$$\begin{cases} \sqrt{2}=(\lambda_1+\lambda_2\cdot 1)\cdot\sqrt{2^1} \\ 8=(\lambda_1+\lambda_2\cdot 2)\cdot\sqrt{2^2} \end{cases}$$

Solving the equation set above and we obtain $\lambda_1=-2$ and $\lambda_2=3$. Substitute them into the equation (a) and the general term becomes

$$a_n=(-2+3\cdot n)\sqrt{2^n}.$$

Example 4.2.19   *Recursive Sequences of Type 5*

Find the general term a_n of the sequence $\{a_n\}$ using two different methods if $a_1 = 2$, $a_2 = 14$, and $a_{n+2}=4a_{n+1}-3a_n$ $(n\geqslant 1)$.

Solution:

- Characteristic Equation Method

  Arrange $$a_{n+2}=4a_{n+1}-3a_n$$

  to $$a_{n+2}-4a_{n+1}+3a_n=0.$$

  The corresponding characteristic equation is

  $$x^2-4\cdot x+3=0.$$

  Since the discriminant

  $$\Delta=(-4)^2-4\cdot 3=4>0,$$

  there are two different real roots $x_1=1$ and $x_2=3$. By the equation (4.2.5), we have the general term

  $$a_n=\lambda_1\cdot 1^n+\lambda_2\cdot 3^n \qquad \text{(a)}$$

  Substitute $a_1=2$, $a_2=14$ for a_n in above equation respectively, we have

  $$\begin{cases} 2=\lambda_1+3^1\cdot\lambda_2 \\ 14=\lambda_1+3^2\cdot\lambda_2 \end{cases}$$

  We obtain $\lambda_1=-4$ and $\lambda_2=2$. Substitute them into (a), we obtain

  $$a_n=-4+2\cdot 3^n.$$

- Difference method

  Arrange $$a_{n+2}=4a_{n+1}-3a_n$$

  to $$a_{n+2}-a_{n+1}=3(a_{n+1}-a_n).$$

  Let $b_n=a_{n+1}-a_n$ $(n\geqslant 1)$, then

$$\frac{b_{n+1}}{b_n} = \frac{a_{n+2} - a_{n+1}}{a_{n+1} - a_n} = 3.$$

Since $b_1 = a_2 - a_1 = 12$, the sequence $\{b_n\}: \{a_{n+1} - a_n\}$ is a geometric sequence whose $b_1 = 12$ and common ratio $r = 3$.

Thus
$$b_n = b_1 \cdot r^{n-1} = 12 \cdot 3^{n-1} = 4\left(3^n\right)$$

and
$$a_{n+1} - a_n = 4\left(3^n\right).$$

Now it is a recursive sequence of Type 1. we list $n - 1$ equations as below.

$$\begin{cases} a_n - a_{n-1} = 4\left(3^{n-1}\right) \\ \cdots \\ a_3 - a_2 - 4\left(3^2\right) \\ a_2 - a_1 = 4(3) \end{cases}$$

Add all equations above.
$$\left(a_n - a_{n-1}\right) + \left(a_{n-1} - a_{n-2}\right) + \cdots + \left(a_2 - a_1\right)$$
$$= 4\left(3^{n-1}\right) + 4\left(3^{n-2}\right) + \cdots + 4(3)$$

Therefore
$$a_n = a_1 + 4 \cdot \left(3^{n-1} + 3^{n-2} + \cdots + 3\right)$$
$$= 2 + 4 \cdot \frac{3\left(1 - 3^{n-1}\right)}{1 - 3}$$
$$= 2\left(3^n\right) - 4$$

▶ Type 6: $a_{n+2} = p \cdot a_{n+1} + q \cdot a_n + c$

If a sequence $\{a_n\}$ is defined by the following recursive formula

$$a_{n+2} = pa_{n+1} + qa_n + c \qquad (n \geqslant 1), \ c \text{ is a constant}$$

and two initials a_1 and a_2, we can convert it to a recursive sequences of Type 5.

As $a_{n+2} = pa_{n+1} + qa_n + c$ and $a_{n+3} = pa_{n+2} + qa_{n+1} + c$,

$$a_{n+3} - a_{n+2} = p(a_{n+2} - a_{n+1}) + q(a_{n+1} - a_n).$$

Let $b_n = a_{n+1} - a_n$, then $b_{n+2} = pb_{n+1} + qb_n \qquad (n \geqslant 1).$

Now the sequence $\{b_n\}$ is a recursive sequences of Type 5 thus we can obtain the general term $b_n = f(n)$ with two initials

$$b_1 = a_2 - a_1$$

and

$$b_2 = a_3 - a_2 = (p-1)a_2 + qa_1 + c.$$

Then

$$b_n = a_{n+1} - a_n = f(n)$$

and it is a recursive sequence of Type 1 now so we can find the general term a_n.

Example 4.2.20   Recursive Sequences of Type 6

If $a_1 = 1$, $a_2 = 3$, convert the recursive sequence

$$a_{n+2} = 5a_{n+1} - 6a_n + 4 \qquad (n \geqslant 1).$$

to a recursive sequence of Type 1.

Solution:

Since

$$a_{n+3} = 5a_{n+2} - 6a_{n+1} + 4,$$
$$a_{n+3} - a_{n+2} = 5(a_{n+2} - a_{n+1}) - 6(a_{n+1} - a_n) \qquad \text{(a)}$$

Let $b_n = a_{n+1} - a_n$, and we have two initials, $b_1 = a_2 - a_1 = 2$ and $b_2 = a_3 - a_2 = 10$ ($a_3 = 5a_2 - 6a_1 + 4 = 13$). Then the formula (a) becomes

$$b_{n+2} - 5b_{n+1} + 6b_n = 0. \qquad \text{(b)}$$

Now $\{b_n\}$ is a recursive sequence of Type 5. Its corresponding characteristic equation is

$$x^2 - 5 \cdot x + 6 = 0.$$

Since

$$\Delta = (-5)^2 - 4 \cdot 6 = 1 > 0,$$

there are two different real roots $x_1 = 3$, $x_2 = 2$.

By the formula (4.2.5) $b_n = \lambda_1 \cdot 3^n + \lambda_2 \cdot 2^n.$ \qquad (c)

Substitute b_1 and b_2 into (c) respectively and we get

$$\begin{cases} 2=\lambda_1 \cdot 3^1 + \lambda_2 \cdot 2^1 \\ 10=\lambda_1 \cdot 3^2 + \lambda_2 \cdot 2^2 \end{cases}.$$

Thus $\lambda_1 = 2$, $\lambda_2 = -2$ and $b_n = (2)3^n - 2^{n+1}$. That is

$$a_{n+1} - a_n = (2) \cdot 3^n - 2^{n+1}.$$

It is a recursive sequence of Type 1 and further we can obtain the general term a_n.

Example 4.2.21   Recursive Sequences of Type 6

Find the general term a_n of the recursive sequence $\{a_n\}$ if $a_1 = 1$, $a_2 = 10$, and $a_{n+2} = 8a_{n+1} - 7a_n - 12 \ (n \geqslant 1)$.

Solution:

Since

$$a_{n+3} = 8a_{n+2} - 7a_{n+1} - 12,$$
$$a_{n+3} - a_{n+2} = 8(a_{n+2} - a_{n+1}) - 7(a_{n+1} - a_n)$$

Let $b_n = a_{n+1} - a_n$, we have $b_{n+2} = 8b_{n+1} - 7b_n$ (a)

It has two initials, $b_1 = a_2 - a_1 = 9$

and $b_2 = a_3 - a_2 = 51$ (here $a_3 = 8a_2 - 7a_1 - 12 = 61$).

Now the sequence $\{b_n\}$ is a recursive sequence of Type 5.

Arrange (a) to $b_{n+2} - 8b_{n+1} + 7b_n = 0.$

The corresponding characteristic equation of the recursive equation (a) is

$$x^2 - 8 \cdot x + 7 = 0.$$

Since

$$\Delta = (-8)^2 - 4 \cdot 7 = 36 > 0,$$

there are two different real roots $x_1 = 7$, $x_2 = 1$. By the (4.2.5), we have the general term

$$b_n = \lambda_1 \cdot 7^n + \lambda_2 \cdot 1^n \qquad (b)$$

Substitute $b_1 = 9$, $b_2 = 51$ for b_n respectively in above equation and obtain the equation set

$$\begin{cases} 9 = 7^1 \cdot \lambda_1 + 1^1 \cdot \lambda_2 \\ 51 = 7^2 \cdot \lambda_1 + 1^2 \cdot \lambda_2 \end{cases}$$

We obtain $\lambda_1 = 1$ and $\lambda_2 = 2$ and substitute them for λ_1 and λ_2 of (b).

$$b_n = 1 \cdot 7^n + 2 \cdot 1^n = 7^n + 2.$$

Since $b_n = a_{n+1} - a_n$, we have $a_{n+1} - a_n = 7^n + 2$. The sequence $\{a_n\}$ is a recursive sequence of Type 1. We have the following $n - 1$ equations.

$$\begin{cases} a_n - a_{n-1} = 7^{n-1} + 2 \\ a_{n-1} - a_{n-2} = 7^{n-2} + 2 \\ \cdots \\ a_3 - a_2 = 7^2 + 2 \\ a_2 - a_1 = 7^1 + 2 \end{cases}$$

Add all of them and

$$a_n - a_1 = 2(n-1) + (7 + 7^2 + \cdots + 7^{n-1}).$$

We obtain
$$a_n = 1 + 2(n-1) + \frac{7(1 - 7^{n-1})}{1 - 7}$$

$$= 2n + \frac{7^n - 13}{6}.$$

Example 4.2.22   Recursive Sequences of Type 6

Find the general term of the sequence $\{a_n\}$ if $a_1 = 2$, $a_2 = 3$, and
$$a_{n+2} = 9a_{n+1} - 18a_n + 3^{n+1} \quad (n \geqslant 1).$$

Solution:

Arrange
$$a_{n+2} = 9a_{n+1} - 18a_n + 3^{n+1}$$

to
$$(a_{n+2} - 3a_{n+1}) - 6(a_{n+1} - 3a_n) = 3^{n+1}.$$

Let $b_n = a_{n+1} - 3a_n$ and substitute it for $a_{n+1} - 3a_n$ of above equation, then

$$b_1 = a_2 - 3a_1 = -3,$$

and
$$b_{n+1} - 6b_n = 3^{n+1} \qquad (n \geqslant 1). \qquad \text{(a)}$$

Divide both sides of (a) by 6^{n+1},

we have
$$\frac{b_{n+1}}{6^{n+1}} - \frac{b_n}{6^n} = \left(\frac{1}{2}\right)^{n+1}.$$

Let $c_n = \frac{b_n}{6^n}$ and substitute it for $\frac{b_n}{6^n}$ of above equation. We have

$$c_{n+1} - c_n = \left(\frac{1}{2}\right)^{n+1} \quad (n \geqslant 1)$$

and
$$c_1 = \frac{b_1}{6^1} = -\frac{1}{2}.$$

It is a recursive sequence of Type 4. We have $n - 1$ equations

$$c_2 - c_1 = \left(\frac{1}{2}\right)^2, \ c_3 - c_2 = \left(\frac{1}{2}\right)^3, \ \cdots, \ c_n - c_{n-1} = \left(\frac{1}{2}\right)^n$$

Add all $n - 1$ equations above

$$\left(c_2 - c_1\right) + \left(c_3 - c_2\right) + \cdots + \left(c_n - c_{n-1}\right) = \left(\frac{1}{2}\right)^2 + \left(\frac{1}{2}\right)^3 + \cdots + \left(\frac{1}{2}\right)^n$$

$$c_n - c_1 = \frac{1}{2} - \left(\frac{1}{2}\right)^n$$

Then
$$c_n = -\left(\frac{1}{2}\right)^n.$$

Since $b_n = 6^n \, c_n = -3^n$,

$$a_{n+1} - 3 a_n = 3^n. \tag{b}$$

It is also a recursive sequence of Type 4. Divide both side of (b) by 3^{n+1},

we have
$$\frac{a_{n+1}}{3^{n+1}} - \frac{a_n}{3^n} = \frac{-3^n}{3^{n+1}}.$$

Let $e_n = \frac{a_n}{3^n}$ and substitute it for $\frac{a_n}{3^n}$ of the equation above.

Then
$$e_{n+1} - e_n = -\frac{1}{3} \ (n > 1)$$

and
$$e_1 = \frac{a_1}{3^1} = \frac{2}{3}.$$

Since $e_{n+1} - e_n$ is a constant, the sequence $\{e_n\}$ is an arithmetic sequence with the first term e_1 and common difference $d = -\frac{1}{3}$.

Then
$$e_n = e_1 + (n-1)\left(\frac{-1}{3}\right) = 1 - \frac{1}{3} \cdot n.$$

Because $a_n = 3^n e_n$, we obtain the general term of the sequence $\{a_n\}$

$$a_n = (3 - n) \cdot 3^{n-1}.$$

 5 Arithmetic Sequences of Higher Order

▮ 5.1 Introduction

Now we are going to extend the definition of arithmetic sequences to a new concept called **arithmetic sequences of higher order**. At first we give the definition of the **difference-sequences for a sequence** $\{a_n\}$.

▶ Difference-Sequence of Order *k* for a Sequence

For a sequence $\{a_n\}$ of real numbers we can construct its difference-sequences as below.

1) Starting from the second term of $\{a_n\}$, we get the following differences
$$b_1 = a_2 - a_1, \quad b_2 = a_3 - a_2, \quad b_3 = a_4 - a_3, \quad \dots, \quad b_n = a_n - a_{n-1}, \quad \dots$$
to construct the sequence $\{b_n\}$ called the difference-sequence of order 1 for $\{a_n\}$.

2) Similarly we can get the following differences from the sequence $\{b_n\}$
$$c_1 = b_2 - b_1, \quad c_2 = b_3 - b_2, \quad c_3 = b_4 - b_3, \quad \dots, \quad c_n = b_n - b_{n-1}, \quad \dots$$
to construct the sequence $\{c_n\}$ which is called
   - the difference-sequence of order 2 for $\{a_n\}$ and
   - the difference-sequence of order 1 for $\{b_n\}$.

3) Further we can get the following differences from the sequence $\{c_n\}$
$$d_1 = c_2 - c_1, \quad d_2 = c_3 - c_2, \quad d_3 = c_4 - c_3, \quad \dots, \quad d_n = c_n - c_{n-1}, \quad \dots$$
to construct the sequence $\{d_n\}$ which is called
   - the difference-sequence of order 3 for $\{a_n\}$,
   - the difference-sequence of order 2 for $\{b_n\}$, and
   - the difference-sequence of order 1 for $\{c_n\}$.

4) Continually we can find the difference-sequence of order k for $\{a_n\}$.

The figure below shows how it works.

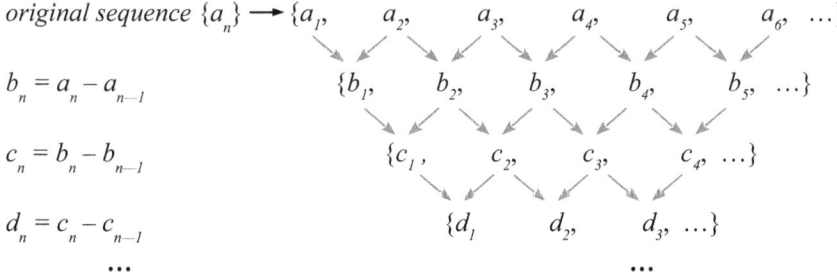

We use $\{\Delta^k a_n\}$ to denote the difference-sequence of order k for a sequence $\{a_n\}$. For instance, the difference-sequence of order 6 for $\{a_n\}$ is denoted by $\{\Delta^6 a_n\}$.

DEFINITION   Difference-Sequence of Order k

Suppose a sequence $\{a_n\}$ is a sequence of real numbers. The sequence $\{\Delta^k a_n\}$ is the difference-sequence of order k for the sequence $\{a_n\}$ if

$$\Delta^k a_n = \Delta^{k-1} a_{n+1} - \Delta^{k-1} a_n \qquad (k \geqslant 1, k \in N^+) \qquad (5.1.1)$$

Where $\Delta^0 a_n$ denotes a_n and $\{\Delta^0 a_n\}$ represents the sequence $\{a_n\}$.

From the definition above, the sequence $\{\Delta^k a_n\}$ is the difference-sequence of order 1 for the difference-sequence $\{\Delta^{k-1} a_n\}$, the difference-sequence of order 2 for the sequence $\Delta^{k-2} a_n$, and so on.

▶ Arithmetic Sequence of Order k

DEFINITION   Arithmetic Sequence of Order k

A sequence $\{a_n\}$ is an arithmetic sequence of order k ($k \geqslant 2$) if its difference-sequence of order k is a nonzero-constant sequence but its difference-sequence of order $k–1$ is not.

If the sequence $\{a_n\}$ is an arithmetic sequence of order 1 then it is a regular arithmetic sequence. In this section we discuss arithmetic sequences of order k ($k \geqslant 2$) only. For simplicity in discussion we define the following abbreviations of some terminologies.
- ASHO   arithmetic sequences of high order
- DS   difference-sequences
- AS   arithmetic sequences

- NZCS   nonzero-constant sequences
- ZCS   zero-constant sequences

Let's look at an example, $\{a_n\}:\{5,9,16,28,47,75,\cdots\}$.

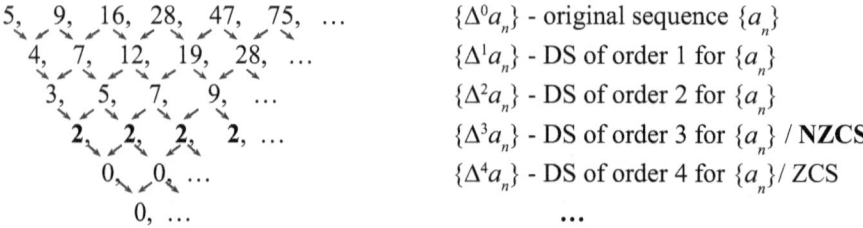

5, 9, 16, 28, 47, 75, ...	$\{\Delta^0 a_n\}$ - original sequence $\{a_n\}$
4, 7, 12, 19, 28, ...	$\{\Delta^1 a_n\}$ - DS of order 1 for $\{a_n\}$
3, 5, 7, 9, ...	$\{\Delta^2 a_n\}$ - DS of order 2 for $\{a_n\}$
2, 2, 2, 2, ...	$\{\Delta^3 a_n\}$ - DS of order 3 for $\{a_n\}$ / **NZCS**
0, 0, ...	$\{\Delta^4 a_n\}$ - DS of order 4 for $\{a_n\}$/ ZCS
0, ...	...

We explain it in detail as below.

1) Let $\Delta^1 a_n = a_{n+1} - a_n$, $\{\Delta^1 a_n\}:\{4, 7, 12, ... \}$ is DS of order 1 for $\{a_n\}$.
2) Let $\Delta^2 a_n = \Delta^1 a_{n+1} - \Delta^1 a_n$, $\{\Delta^2 a_n\}:\{3, 5, 7, ... \}$ is DS of order 2 for $\{a_n\}$.
3) Let $\Delta^3 a_n = \Delta^2 a_{n+1} - \Delta^2 a_n$, $\{\Delta^3 a_n\}:\{2, 2, 2, ... \}$ is DS of order 3 for $\{a_n\}$.
4) All difference-sequences of order k $(k>3)$ are zero-constant sequences $\{0, 0, ...\}$.

The difference-sequence of order 3 for the sequence $\{a_n\}$, $\{\Delta^3 a_n\}$, is a nonzero-constant sequence $\{2, 2, ...\}$ and all the difference-sequences of order k $(k>3)$ are zero-constant sequences $\{0, 0, ...\}$, thus the sequence $\{a_n\}$ is an arithmetic sequence of order 3.

THEOREM 5.1.1

If a sequence $\{a_n\}$ is an arithmetic sequence of order K $(K \geqslant 2)$, its difference-sequence of order k $(k<K)$ is the arithmetic sequence of order $(K - k)$.

It is straightforward to explain the theorem 5.1.1 by previous example. We know the sequence $\{a_n\}$ is the arithmetic sequence of order 3 $(K = 3)$.

5, 9, 16, 28, 47, ...	$\{\Delta^0 a\}:\{a_n\}$ and AS of order 3
4, 7, 12, 19, ...	$\{\Delta^1 a_n\}$ - DS of order 1 (k) and AS of order 2 (3-1)
3, 5, 7, ...	$\{\Delta^2 a_n\}$ - DS of order 2 (k) and AS of order 1 (3-2)
2, 2, 2, ...	$\{\Delta^3 a_n\}$ - DS of order 3 (K) / **NZCS**
0, 0, ...	$\{\Delta^4 a_n\}$ - DS of order 4 / ZCS
0, ...	...

Example 5.1.1   *Arithmetic Sequences of Higher Order*

Show the order of arithmetic sequence of higher order for the following sequences.

1) $\{a_n\}:\{9,16,28,48,79,124,\cdots\}$
2) $\{a_n\}:\{2,5,15,36,74,142,265,\cdots\}$

Solution:

1)

9, 16, 28, 48, 79, 124, ...	$\{\Delta^0 a_n\}:\{a_n\}$
7, 12, 20, 31, 45, ...	$\{\Delta^1 a_n\}$ - DS of order 1 for $\{a_n\}$
5, 8, 11, 14, ...	$\{\Delta^2 a_n\}$ - DS of order 2 for $\{a_n\}$
3, 3, 3, ...	$\{\Delta^3 a_n\}$ - DS of order 3 for $\{a_n\}$ / NZCS

The sequence $\{a_n\}$ is an arithmetic sequence of order 3.

2)

2, 5, 15, 36, 74, 142, 265, ...	$\{\Delta^0 a_n\}:\{a_n\}$
3, 10, 21, 38, 68, 123, ...	$\{\Delta^1 a_n\}$ - DS of order 1 for $\{a_n\}$
7, 11, 17, 30, 55, ...	$\{\Delta^2 a_n\}$ - DS of order 2 for $\{a_n\}$
4, 6, 13, 25, ...	$\{\Delta^3 a_n\}$ - DS of order 3 for $\{a_n\}$
2, 7, 12, ...	$\{\Delta^4 a_n\}$ - DS of order 4 for $\{a_n\}$
5, 5, 5, ...	$\{\Delta^5 a_n\}$ - DS of order 5 for $\{a_n\}$ / NZCS

Therefore the sequence $\{a_n\}$ is an arithmetic sequence of order 5.

■ 5.2 General Term

Suppose a sequence $\{a_n\}$ is an arithmetic sequence of order k $(k \geqslant 2)$.

$$
\begin{array}{llllllll}
a_1, & a_2, & a_3, & a_4, & a_5, & a_6, & \ldots, & a_n, \ldots & \{\Delta^0 a_n\} = \{a_n\} \\
\Delta^1 a_1, & \Delta^1 a_2, & \Delta^1 a_3, & \Delta^1 a_4, & \Delta^1 a_5, & \Delta^1 a_6, & \ldots & & \{\Delta^1 a_n\} \\
& \Delta^2 a_1, & \Delta^2 a_2, & \Delta^2 a_3, & \Delta^2 a_4, & \Delta^2 a_5, & \ldots & & \{\Delta^2 a_n\} \\
& & \Delta^3 a_1, & \Delta^3 a_2, & \Delta^3 a_3, & \Delta^3 a_2, & \ldots & & \{\Delta^3 a_n\} \\
& & & \cdots & \cdots & \cdots & & & \cdots \\
& & & \Delta^k a_1, & \Delta^k a_2, & \Delta^k a_3, & \ldots & & \{\Delta^k a_n\} \text{ is NZCS} \\
& & & & 0, & 0, & \ldots & & \{\Delta^{k+1} a_n\} = \{0\}
\end{array}
$$

From above we have the followings.

$$a_2 - a_1 = \Delta^1 a_1$$

$$a_3 - a_2 = \Delta^1 a_2 = \Delta^1 a_1 + \Delta^2 a_1 = \binom{1}{0}\Delta^1 a_1 + \binom{1}{1}\Delta^2 a_1$$

$$a_4 - a_3 = \Delta^1 a_3 = \Delta^1 a_2 + \Delta^2 a_2 = \Delta^1 a_1 + 2\Delta^2 a_1 + \Delta^3 a_1 = \binom{2}{0}\Delta^1 a_1 + \binom{2}{1}\Delta^2 a_1 + \binom{2}{2}\Delta^3 a_1$$

$$a_5 - a_4 = \Delta^1 a_4 = \Delta^1 a_3 + \Delta^2 a_3 = \Delta^1 a_2 + 2\Delta^2 a_2 + \Delta^3 a_2$$

$$= (\Delta^1 a_1 + \Delta^2 a_1) + 2(\Delta^2 a_1 + \Delta^3 a_1) + (\Delta^3 a_1 + \Delta^4 a_1)$$

$$= \Delta^1 a_1 + 3\Delta^2 a_1 + 3\Delta^3 a_1 + \Delta^4 a_1 = \binom{3}{0}\Delta^1 a_1 + \binom{3}{1}\Delta^2 a_1 + \binom{3}{2}\Delta^3 a_1 + \binom{3}{3}\Delta^4 a_1$$

$$\cdots$$

$$a_n - a_{n-1} = \binom{n-2}{0}\Delta^1 a_1 + \binom{n-2}{1}\Delta^2 a_1 + \binom{n-2}{2}\Delta^3 a_1 + \cdots + \binom{n-2}{n-2}\Delta^{n-1} a_1.$$

Because the sequence $\{a_n\}$ is an arithmetic sequence of order k $(k \geqslant 2)$, its difference-sequence of order k, the sequence $\{\Delta^k a_n\}$, must be a nonzero-constant sequence.

$$a_n - a_{n-1} = \binom{n-2}{0}\Delta^1 a_1 + \binom{n-2}{1}\Delta^2 a_1 + \binom{n-2}{2}\Delta^3 a_1 + \cdots + \binom{n-2}{k-1}\Delta^k a_1.$$

On the other hand, we have

$$a_n = a_1 + (a_2 - a_1) + (a_3 - a_2) + \cdots + (a_n - a_{n-1})$$

$$= a_1 + \Delta^1 a_1$$

$$+ \binom{1}{0}\Delta^1 a_1 + \binom{1}{1}\Delta^2 a_1$$

$$+ \binom{2}{0}\Delta^1 a_1 + \binom{2}{1}\Delta^2 a_1 + \binom{2}{2}\Delta^3 a_1$$

$$+ \binom{3}{0}\Delta^1 a_1 + \binom{3}{1}\Delta^2 a_1 + \binom{3}{2}\Delta^3 a_1 + \binom{3}{3}\Delta^4 a_1$$

$$\cdots \cdots$$

$$+ \binom{n-2}{0}\Delta^1 a_1 + \binom{n-2}{1}\Delta^2 a_1 + \binom{n-2}{2}\Delta^3 a_1 + \binom{n-2}{3}\Delta^4 a_1 + \cdots + \binom{n-2}{k-1}\Delta^k a_1$$

Notice the columns in above equation, we can arrange the equation above like

$$
a_n = \left.
\begin{aligned}
& a_1 \\
&+ \left(\binom{0}{0}+\binom{1}{0}+\binom{2}{0}+\cdots+\binom{n-2}{0}\right)\Delta^1 a_1 \\
&+ \left(\binom{1}{1}+\binom{2}{1}+\binom{3}{1}+\cdots+\binom{n-2}{1}\right)\Delta^2 a_1 \\
&+ \left(\binom{2}{2}+\binom{3}{2}+\binom{4}{2}+\cdots+\binom{n-2}{2}\right)\Delta^3 a_1 \\
&\qquad\cdots \\
&+ \left(\binom{k-1}{k-1}+\binom{k}{k-1}+\binom{k+1}{k-1}+\cdots+\binom{n-2}{k-1}\right)\Delta^k a_1 .
\end{aligned}
\right\}
\quad (a)
$$

Substitute $n-2$ and $k-1$ for n and m of (b) and apply the identity of combinations (b) below to each term of the equation (a).

$$\binom{m}{m}+\binom{m+1}{m}+\binom{m+2}{m}+\cdots+\binom{n}{m}=\binom{n+1}{m+1} \qquad (b)$$

Then the terms in (a) become

the 1st term: $\qquad\qquad\qquad\qquad a_1=\binom{n-1}{0}a_1$

the 2nd term ($m=0,\ i=n-1$): $\qquad\qquad \binom{n-1}{1}\Delta^1 a_1$

the 3rd term ($m=1,\ i=n-2$): $\qquad\qquad \binom{n-1}{2}\Delta^2 a_1$

$$\cdots \cdots$$

the kth term ($m=k-1, i=n-k$): $\qquad \binom{n-1}{k}\Delta^k a_1$

Then the general term a_n can be expressed as below.

The General Term of Arithmetic Sequences of Order k

If $\{a_n\}$ is an arithmetic sequence of order k ($k\geqslant 2$), its general term a_n is

$$u_n-\binom{n-1}{0}u_1+\binom{n-1}{1}\Delta^1 a_1+\binom{n-1}{2}\Delta^2 a_1+\cdots+\binom{n-1}{k}\Delta^k a_1 \quad (5.2.1)$$

THEOREM 5.2.1

A sequence $\{a_n\}$ is an arithmetic sequence of order k $(k \geqslant 2)$ if and only if the general term a_n of $\{a_n\}$ is a polynomial of degree k in terms of n.

$$a_n = \lambda_k n^k + \lambda_{k-1} n^{k-1} + \cdots + \lambda_2 n^2 + \lambda_1 n^1 + \lambda_0 \qquad (5.2.2)$$

Where $\lambda_0, \lambda_1, \lambda_2, \ldots, \lambda_k$ are constants.

This theorem can be used to identify arithmetic sequences of higher order. To help the readers to understand the identity of combinations (b) above we give the proofs of the following identities of combinations.

1) $\dbinom{m}{m} = \dbinom{m+1}{m+1}$

Proof   *Identity of Combinations*

Since
$$\binom{m}{m} = \frac{m!}{(m-m)!\,m!} = \frac{m!}{0!\,m!} = 1$$

and
$$\binom{m+1}{m+1} = \frac{(m+1)!}{[(m+1)-(m+1)]!\,(m+1)!} = \frac{(m+1)!}{0!\,(m+1)!} = 1,$$

we have
$$\binom{m}{m} = \binom{m+1}{m+1}.$$

2) $\dbinom{n}{m+1} + \dbinom{n}{m} = \dbinom{n+1}{m+1}$ $(n \geqslant m)$

Proof   *Identity of Combinations*

Since
$$\binom{n}{m+1} + \binom{n}{m} = \frac{n!}{(n-m-1)!\,(m+1)!} + \frac{n!}{(n-m)!\,m!}$$
$$= \frac{n!(n-m)}{(n-m)!\,(m+1)!} + \frac{n!(m+1)}{(n-m)!\,(m+1)!}$$
$$= \frac{(n+1)!}{(n-m)!\,(m+1)!} = \binom{n+1}{m+1}.$$

Then
$$\binom{n}{m+1} + \binom{n}{m} = \binom{n+1}{m+1}.$$

3) $\dbinom{m}{m} + \dbinom{m+1}{m} + \dbinom{m+2}{m} + \cdots + \dbinom{n}{m} = \dbinom{n+1}{m+1}$ $(n \geqslant m)$

Proof　　　　Identity of Combinations

Since $\binom{m}{m}=\binom{m+1}{m+1}$, the left side of the equation becomes

$$\binom{m+1}{m+1}+\binom{m+1}{m}+\binom{m+2}{m}+\cdots+\binom{n}{m} \qquad (a)$$

As

$$\binom{n}{m+1}+\binom{n}{m}=\binom{n+1}{m+1} \qquad (b)$$

we apply the identity (b) to the first two items of (a). Then (a) becomes

$$\binom{m+2}{m+1}+\binom{m+2}{m}+\binom{m+3}{m}+\binom{m+4}{m}+\cdots+\binom{n}{m} \qquad (c)$$

Apply equation (b) repeatedly to the first two items of (c), then

$$\binom{m}{m}+\binom{m+1}{m}+\binom{m+2}{m}+\cdots+\binom{n}{m}=\binom{n+1}{m+1}.$$

Example 5.2.1　　General Term of ASHO

Find the general term a_n of the sequence $\{9, 16, 28, 48, 79, 124, \cdots\}$.

Solution:

List each difference-sequence of $\{a_n\}$ until reach a nonzero-constant sequence.

9,	16,	28,	48,	79,	124,	...	$\{\Delta^0 a_n\}:\{a_n\}$
	7,	12,	20,	31,	45,	...	$\{\Delta^1 a_n\}$
		5,	8,	11,	14,	...	$\{\Delta^2 a_n\}$
			3,	3,	3,	...	$\{\Delta^3 a_n\}$ / NZCS

Then sequence $\{a_n\}$ is an arithmetic sequence of order 3.

$$a_n=\binom{n-1}{0}\cdot 9+\binom{n-1}{1}\cdot 7+\binom{n-1}{2}\cdot 5+\binom{n-1}{3}\cdot 3=\frac{1}{2}\cdot n^3-\frac{1}{2}\cdot n^2+5\cdot n+4.$$

Example 5.2.2　　General Term of ASHO

Find the general term a_n of the sequence $\{10, 14, 23, 37, 56, 80, \cdots\}$.

Solution:

List each difference-sequence of $\{a_n\}$ until reach nonzero-constant sequence.

10,	14,	23,	37,	56,	80,	...	$\{\Delta^0 a_n\}:\{a_n\}$
	4,	9,	14,	19,	24,	...	$\{\Delta^1 a_n\}$
		5,	5,	5,	...		$\{\Delta^2 a_n\}$ / NZCS

Then sequence $\{a_n\}$ is an arithmetic sequence of order 2.

$$a_n=\binom{n-1}{0}\cdot 10+\binom{n-1}{1}\cdot 4+\binom{n-1}{2}\cdot 5=\frac{5}{2}\cdot n^2-\frac{7}{2}\cdot n+11.$$

Example 5.2.3   *General Term of ASHO*

Let the difference-sequence of order 2 for a sequence $\{a_n\}$ be a constant sequence, $\{\Delta^2 a_n\}:\{3, 3, 3, \ldots\}$, $a_4=30$, and $a_8=100$. Find the general term a_n.

Solution:

The general term a_n becomes

$$a_n=\binom{n-1}{0}\cdot a_1+\binom{n-1}{1}\cdot\Delta^1 a_1+\binom{n-1}{2}\cdot\Delta^2 a_1 \tag{a}$$

Since $\Delta^2 a_n = 3$, $a_4=30$ and $a_8=100$ we have the equation set as below.

$$\begin{cases} a_4=\binom{4-1}{0}\cdot a_1+\binom{4-1}{1}\cdot\Delta^1 a_1+\binom{4-1}{2}\cdot 3=30 \\ a_8=\binom{8-1}{0}\cdot a_1+\binom{8-1}{1}\cdot\Delta^1 a_1+\binom{8-1}{2}\cdot 3=100 \end{cases}$$

Solve the equation set above, we get $a_1=9$ and $\Delta^1 a_1 = 4$. Substitute them into (a) then

$$a_n=\binom{n-1}{0}\cdot 9+\binom{n-1}{1}\cdot 4+\binom{n-1}{2}\cdot 3=\frac{3}{2}\cdot n^2-\frac{1}{2}\cdot n+8.$$

Example 5.2.4   *General Term of ASHO*

The difference-sequence of order 2 for a sequence $\{a_n\}$ is a geometric sequence and the first four terms of $\{a_n\}$ are 2, 7, 15, and 32. Find the general term a_n.

Solution:

List the first two difference-sequences.

2,	7,	15,	32,	...
	5,	8,	17,	...
		3,	9,	...

$\{\Delta^0 a_n\}:\{a_n\}$

$\{\Delta^1 a_n\}$

$\{\Delta^2 a_n\}$ is a geometric sequence, $r = 3$

Let's use the following graph to explain.

$a_1=2,\quad a_2=7,\quad a_3=15,\quad \ldots,\quad a_n,\quad a_{n+1},\quad a_{n+2}, \ldots \qquad \{\Delta^0 a_n\}:\{a_n\}$

$\Delta^1 a_1=5,\quad \Delta^1 a_2=8,\quad \Delta^1 a_3=17,\ldots,\quad \Delta^1 a_n,\quad \Delta^1 a_{n+1}, \ldots \qquad \{\Delta^1 a_n\}$

$\Delta^2 a_1=3,\quad \Delta^2 a_2=9,\quad \ldots,\quad \Delta^2 a_n, \ldots \qquad \{\Delta^2 a_n\}$

Because the difference-sequence of order 2, $\{\Delta^2 a_n\}$, is a geometric sequence, $\Delta^2 a_1 = 3$, and common ratio $r=3$, then the general term $\Delta^2 a_n=\Delta^2 a_1\cdot 3^{n-1}=3^n$. We have

$$\Delta^1 a_{n+1}-\Delta^1 a_n=\Delta^2 a_n=3^n.$$

It is a recursive sequence of Type 1. Then we have $n - 1$ equations

$$\begin{cases} \Delta^1 a_n - \Delta^1 a_{n-1} = 3^{n-1} \\ \Delta^1 a_{n-1} - \Delta^1 a_{n-2} = 3^{n-2} \\ \cdots \\ \Delta^1 a_2 - \Delta^1 a_1 = 3 \end{cases}$$

Add all $n - 1$ equations.

$$\Delta^1 a_n - \Delta^1 a_1 = 3 + 3^2 + 3^3 + \cdots + 3^{n-1}.$$
$$= \frac{3(1-3^{n-1})}{1-3} = \frac{(3^n-3)}{2}$$

Since $\Delta^1 a_1 = 5$, then $\qquad \Delta^1 a_n = 5 + \dfrac{(3^n-3)}{2} = \dfrac{3^n}{2} + \dfrac{7}{2}.$

Because $a_{n+1} - a_n = \Delta^1 a_n$,

$$a_{n+1} - a_n = \frac{3^n}{2} + \frac{7}{2}$$

It is a recursive sequence of type 1. List the following $n - 1$ equations

$$\begin{cases} a_n - a_{n-1} = \dfrac{3^{n-1}}{2} + \dfrac{7}{2} \\ \cdots \\ a_2 - a_1 = \dfrac{3}{2} + \dfrac{7}{2} \end{cases}$$

Add all $n - 1$ equations above.

$$a_n - a_1 = \frac{1}{2}(3 + 3^2 + \cdots + 3^{n-1}) + \frac{7}{2}(n-1)$$

Because $\qquad 3 + 3^2 + 3^3 + \cdots + 3^{n-1} = \dfrac{3(1-3^{n-1})}{1-3} = \dfrac{3^n}{2} - \dfrac{3}{2}$

and $a_1 = 2$, we have

$$a_n = \frac{1}{4}(3^n + 14n - 9).$$

■ 5.3 The Sum of the First n Terms

Suppose $\{a_n\}$ is an arithmetic sequence of order k.

$$\{a_1, a_2, a_{3,} \cdots, a_{n-1}, a_n, \cdots\}$$

The sum of the first n terms of the sequence $\{a_n\}$ is

$$S_n = a_1 + a_2 + \cdots + a_n$$

Let's look at the sequence $\{S_n\}$ consisting of the sums as below.

$$\{0, S_1, S_2, S_{3,} \cdots, S_{n-1}, S_n, \cdots\}$$

Because $S_1 - 0 = a_1$, $S_2 - S_1 = a_2$, ... , $S_n - S_{n-1} = a_n$, the sequence $\{S_n\}$ is an arithmetic sequence of order $k+1$ by the theorem 5.1.1. It is straightforward when you put the sequence $\{S_n\}$ on the top of $\{a_n\}$ like below.

$0,$	$S_1,$	$S_2,$	$S_3,$	...,	$S_{n-1},$	$S_n,$ ...	$\{S_n\}$ is AS of order $k+1$
$a_1,$	$a_2,$	$a_3,$	...,		$a_n,$ ...		$\{\Delta^0 a_n\} = \{a_n\}$ is AS of order k
	$\Delta^1 a_1,$	$\Delta^1 a_2,$	$\Delta^1 a_3,$	$\Delta^1 a_4,$	...		$\{\Delta^1 a_n\}$
		...		...			...
	$\Delta^k a_1,$	$\Delta^k a_2,$	$\Delta^k a_3,$	...			$\{\Delta^k a_n\}$ / NZCS
		$0,$	$0,$ ...				$\{\Delta^{k+1} a_n\}:\{0, 0, \ldots\}$ / ZCS

We notice that S_n is the $(n+1)^{\text{th}}$ term of sequence $\{S_n\}$, an arithmetic sequence of order $k+1$. By the formula (5.2.1), we obtain the following formula.

Sum of the First n Terms of Arithmetic Sequences of Order k

If $\{a_n\}$ is an arithmetic sequence of order k $(k \geqslant 2)$, the sum S_n of the first n terms of the sequence becomes

$$S_n = \binom{n}{1} a_1 + \binom{n}{2} \Delta^1 a_1 + \binom{n}{3} \Delta^2 a_1 + \cdots + \binom{n}{k+1} \Delta^k a_1. \qquad (5.3.1)$$

THEOREM 5.3.1

If $\{a_n\}$ is an arithmetic sequence of order k $(k \geqslant 2)$ then the sum S_n of the first n terms of the sequence is a polynomial of degree $k+1$ in terms of n.

$$S_n = \lambda_{k+1} n^{k+1} + \lambda_k n^k + \cdots + \lambda_2 n^2 + \lambda_1 n^1 + \lambda_0 \qquad (5.3.2)$$

Where $\lambda_0, \lambda_1, \cdots, \lambda_k, \lambda_{k+1}$ are constants.

Example 5.3.1   The Sum of the First n Terms of an ASHO

The sequence $\{a_n\}: \{7, 13, 23, 37, 55, 77, \cdots\}$ is given. Find S_n and the order k of $\{a_n\}$ if the sequence $\{0, S_1, S_2, \ldots, S_n, \ldots\}$ is an arithmetic sequence of order k.

Solution:

Write each order of difference-sequence until to reach a nonzero-constant sequence.

7,	13,	23,	37,	55,	77,	...
6,	10,	14,	18,	22,		...
4,	4,	4,	4,			...

$\{\Delta^0 a_n\}: \{a_n\}$

$\{\Delta^1 a_n\}$ is the DS of order 1 for $\{a_n\}$

$\{\Delta^2 a_n\}$ is the DS of order 2 for $\{a_n\}$ / NZCS

Then $\{a_n\}$ is an arithmetic sequence of order 2, $a_1 = 7, \Delta^1 a_1 = 6, \Delta^2 a_1 = 4 \cdot$ By the formula (5.3.1) we have

$$S_n = \binom{n}{1} \cdot 7 + \binom{n}{2} \cdot 6 + \binom{n}{3} \cdot 4 = \frac{2}{3} n^3 + n^2 + \frac{16}{3} n$$

Then we list the terms of the sequence $\{S_n\}$ below.

$$\begin{cases} S_1 = \frac{2}{3} \cdot 1^3 + 1^2 + \frac{16}{3} \cdot 1 = 7 \\ S_2 = \frac{2}{3} \cdot 2^3 + 2^2 + \frac{16}{3} \cdot 2 = 20 \\ S_3 = \frac{2}{3} \cdot 3^3 + 3^2 + \frac{16}{3} \cdot 3 = 43 \end{cases}$$

$\cdots$

Notice $\{S_n\}$ and $\{a_n\}$,

$$\{S_n\}: \{0, \quad 7, \quad 20, \quad 43, \quad 80, \quad \ldots \}$$
$$\{a_n\}: \{ \quad 7, \quad 13, \quad 23, \quad 37, \ldots \},$$

Since $a_n = S_{n+1} - S_n$, $\{a_n\}$ is the difference-sequence of order 1 of $\{S_n\}$. As $\{a_n\}$ is an arithmetic sequence of order 2, $\{S_n\}$ is an arithmetic sequence of order 3.

Example 5.3.2   The Sum of the First n Terms of an ASHO

If $\{a_n\}: \{10, 19, 34, 59, 98, 155, \cdots\}$ find S_{10}.

Solution:

Write out each order of difference-sequence until to reach a nonzero-constant sequence.

10,	19,	34,	59,	98,	155,	...
9,	15,	25,	39,	57,		...
6,	10,	14,	18,			...
4,	4,	4,				...

$\{\Delta^0 a_n\}: \{a_n\}$

$\{\Delta^1 a_n\}$ is the DS of order 1 for $\{a_n\}$

$\{\Delta^2 a_n\}$ is the DS of order 2 for $\{a_n\}$

$\{\Delta^3 a_n\}$ is the DS of order 3 for $\{a_n\}$ / NZCS

Then $\{a_n\}$ is an arithmetic sequence of order 3,

$$a_1=10, \Delta^1 a_1=9, \Delta^2 a_1=6, \Delta^3 a_1=4.$$

By the formula (5.3.1),

$$S_n=\binom{n}{1}\cdot 10+\binom{n}{2}\cdot 9+\binom{n}{3}\cdot 6+\binom{n}{4}\cdot 4 = \frac{1}{6}(n^4+20\cdot n^2+39\cdot n)$$

$$S_{10}=\frac{1}{6}(10^4+20\cdot 10^2+39\cdot 10)=2065.$$

Example 5.3.3   The Sum of the First n Terms of an ASHO

Prove that $1^3+2^3+3^3+4^3+\cdots+n^3=(1+2+3+\cdots+n)^2.$

Solution:

1,	8,	27,	64,	125,	216,	...	$\{\Delta^0 a_n\}:\{a_n\}$
	7,	19,	37,	61,	91,	...	$\{\Delta^1 a_n\}$ is the DS of order 1 for $\{a_n\}$
		12,	18,	24,	30,	...	$\{\Delta^2 a_n\}$ is the DS of order 2 for $\{a_n\}$
			6,	6,	6,	...	$\{\Delta^3 a_n\}$ is the DS of order 3 for $\{a_n\}$ / NZCS

Then $\{a_n\}$ is an arithmetic sequence of order 3, $a_1 = 1$, $\Delta^1 a_1 = 7$, $\Delta^2 a_1 = 12$, and $\Delta^3 a_n = 6$. By the formula (5.3.1),

$$S_n=\binom{n}{1}\cdot 1+\binom{n}{2}\cdot 7+\binom{n}{3}\cdot 12+\binom{n}{4}\cdot 6 = \frac{1}{4}(n^4+2\cdot n^3+n^2).$$

Because

$$\frac{1}{4}(n^4+2\cdot n^3+n^2)=(\frac{n(n+1)}{2})^2$$

and

$$1+2+3+\cdots+n=\frac{n(n+1)}{2},$$

we have

$$1^3+2^3+3^3+4^3+\cdots+n^3=(1+2+3+\cdots+n)^2.$$

Example 5.3.4   The Sum of the First n Terms of an ASHO

1) Let $\{a_n\}:\{9,16,28,48,79,124,186,\cdots\}$, find S_n.
2) Find k if the sequence $\{0,S_1,S_2,\cdots S_n,\cdots\}$ is an arithmetic sequence of order k.

Solution:

9,	16,	28,	48,	79,	124,	186,	...	$\{\Delta^0 a_n\}:\{a_n\}$
	7,	12,	20,	31,	45,	62,	...	$\{\Delta^1 a_n\}$ is the DS of order 1 for $\{a_n\}$
		5,	8,	11,	14,	17,	...	$\{\Delta^2 a_n\}$ is the DS of order 2 for $\{a_n\}$
			3,	3,	3,	3,	...	$\{\Delta^3 a_n\}$ is the DS of order 3 for $\{a_n\}$ / NZCS

Then $\{a_n\}$ is an arithmetic sequence of order 3, $a_1 = 9$, $\Delta^1 a_1 = 7$, $\Delta^2 a_1 = 5$, and

$\Delta^3 a_1 = 3$.

1) By the formula (5.3.1) we have

$$S_n = \binom{n}{1} \cdot 9 + \binom{n}{2} \cdot 7 + \binom{n}{3} \cdot 5 + \binom{n}{4} \cdot 3 = \frac{1}{8} n^4 + \frac{1}{12} n^3 + \frac{57}{24} n^2 + \frac{77}{12} n.$$

2) Then we list first few terms of the sequence $\{S_n\}$ below.

$$S_1 = \frac{1}{8} \cdot 1^4 + \frac{1}{12} \cdot 1^3 + \frac{57}{24} \cdot 1^2 + \frac{77}{12} \cdot 1 = 9$$

$$S_2 = \frac{1}{8} \cdot 2^4 + \frac{1}{12} \cdot 2^3 + \frac{57}{24} \cdot 2^2 + \frac{77}{12} \cdot 2 = 25$$

$$\cdots$$

Comparing the sequence $\{S_n\}$ with the sequence $\{a_n\}$.

$$\{S_n\} = \{0, \quad 9, \quad 25, \quad 53, \quad \ldots \}$$

$$\{a_n\} = \{ \quad 9, \quad 16, \quad 28, \quad 48, \ldots \}$$

Since $a_n = S_{n+1} - S_n$, the sequence $\{a_n\}$ is the difference-sequence of order 1 for $\{S_n\}$. Because the sequence $\{a_n\}$ is an arithmetic sequence of order 3, $\{S_n\}$ is an arithmetic sequence of order $k = 4$.

Example 5.3.5   The Sum of the First n Terms of an ASHO

Balls are stacked in a four sided pyramid of 7 layers and every layer has one more ball on each side than that on the layer above. On the top layer there are 2×4 balls in a rectangle. Find the number of balls on 4th layer and total number of balls.

Solution:

8   15   24

There are $(n+1)(n+3)$ balls on the nth layer.

Let $\{a_n\}:\{8, 15, 24, 35, \ldots\}$.

8,	15,	24,	35,	48,	63,	$\ldots$	$\{\Delta^0 a_n\}:\{a_n\}$
	7,	9,	11,	13,	15,	$\ldots$	$\{\Delta^1 a_n\}$ is the DS of order 1 for $\{a_n\}$
		2,	2,	2,	2,	$\ldots$	$\{\Delta^2 a_n\}$ is the DS of order 2 for $\{u_n\}$ / NZCS

The sequence $\{a_n\}$ is an arithmetic sequence of order 2, $a_1 = 8$, $\Delta^1 a_1 = 7$, and $\Delta^2 a_1 = 2$. By the formula (5.2.1),

$$a_n = \binom{n-1}{0} \cdot 8 + \binom{n-1}{1} \cdot 7 + \binom{n-1}{2} \cdot 2 = n^2 + 4n + 3.$$

The number of balls on the 4th layer is $a_4 = 4^2 + 4 \cdot 4 + 3 = 35$.

Total number of balls of n layers

$$S_n=\binom{n}{1}\cdot8+\binom{n}{2}\cdot7+\binom{n}{3}\cdot2=\frac{1}{3}n^3+\frac{5}{2}n^2+\frac{31}{6}n.$$

Then the total number of balls of all 7 layers is

$$S_7=\frac{1}{3}\cdot7^3+\frac{5}{2}\cdot7^2+\frac{31}{6}\cdot7=273.$$

Example 5.3.6   The Sum of the First n Terms of an ASHO

Balls are stacked in a tetrahedron and each layer forms a isosceles triangle. There are three balls on the top layer and one more ball on each side of each layer than that on the layer above. Find the number of the balls on the nth layer and total number of balls.

Solution:

From top to bottom, we list the number of balls on the first few layers.

Let $\{a_n\}:\{3, 6, 10, 15, ...\}$.

$$3 \qquad 6 \qquad 10 \qquad \cdots$$

3,   6,   10,   15,   21,   28,   ... $\{\Delta^0a_n\} : \{a_n\}$
  3,   4,   5,   6,   7, ... $\{\Delta^1a_n\}$ is the DS of order 1 for $\{a_n\}$
    1,   1,   1,   1,   ... $\{\Delta^2a_n\}$ is the DS of order 2 for $\{a_n\}$ / NZCS

The sequence $\{a_n\}$ is an arithmetic sequence of order 2 and $a_1 = 3$, $\Delta^1a_1 = 3$, and $\Delta^2a_1 = 1$.

The number of the balls on the nth layer.

$$a_n=\binom{n-1}{0}\cdot3+\binom{n-1}{1}\cdot3+\binom{n-1}{2}\cdot1=\frac{n^2+3n+2}{2}.$$

Total number of balls

$$S_n=\binom{n}{1}\cdot3+\binom{n}{2}\cdot3+\binom{n}{3}\cdot1=\frac{n^3}{6}+n^2+\frac{11}{6}n.$$

www.ingramcontent.com/pod-product-compliance
Lightning Source LLC
Chambersburg PA
CBHW051534170526
45165CB00002B/727